FROM SQUAW TIT TO WHOREHOUSE MEADOW

FROM SQUAW TIT TO WHOREHOUSE MEADOW

How Maps Name, Claim, and Inflame

MARK MONMONIER

The University of Chicago Press

Chicago and London

The University of Chicago Press, Chicago 60637
The University of Chicago Press, Ltd., London
© Mark Monmonier 2006
All rights reserved. Published 2006
Paperback edition 2007
Printed in the United States of America

16 15 14 13 12 11 10 09 08 07   2 3 4 5 6

ISBN-13: 978-0-226-53465-7 (cloth)
ISBN-13: 978-0-226-53466-4 (paper)
ISBN-10: 0-226-53465-0 (cloth)
ISBN-10: 0-226-53466-9 (paper)

Library of Congress Cataloging-in-Publication Data

Monmonier, Mark S.
    From Squaw Tit to Whorehouse Meadow : how maps name, claim, and inflame / Mark
    Monmonier.
        p.   cm.
    Includes bibliographical references (p.) and index.
    ISBN 0-226-53465-0 (alk. paper)
        1. Names, Geographical—United States—Etymology. 2. Names, Geographical—Etymol-
    ogy. 3. Toponymy. 4. English   language—Etymology—Names. 5. English   language—
    Obscene words. 6. Words, Obscene. I. Title.
        G105.M66 2006
        910'.01'4—dc22

                                                                            2005014683

⊚ The paper used in this publication meets the minimum requirements of the American
National Standard for Information Sciences–Permanence of Paper for Printed Library Materials, ANSI Z39.48-1992.

In memory of David Woodward (1942–2004),
whose writing attracted me to the history of cartography and
whose encouragement and insight I deeply miss

Contents

Preface

My original title for this book, before it was changed, was "Fighting Words," which better describes the story of inflammatory ethnic insults embedded in the cartographic labels of places and geographic features. It's also a tale of power and compromise arising from the mapmaker's pursuit of an orderly process for naming and renaming that avoids confusion, preserves history, and serves diverse political aims. Mapmakers and names scholars call this process applied toponymy—*toponym* means place name, and *toponymy* refers to the study of geographic names. As my chapters show, it's a conservative process, frustratingly slow at times, that evolved to capture local usage, resist confusing duplication and rampant commemorative naming, standardize syntax, and remove abusive nomenclature from government maps. Influenced by the standardization imperative that emerged at the end of the nineteenth century and gathered momentum as maps became more detailed, applied toponymy is an important part of map history. In addition to a significant yet largely unsung role in cartography's adoption and adaptation of electronic technology, its sometimes-fretful response to increased ethnic sensitivity in the late twentieth century richly reflects the map's role as a mediated portrait of the cultural landscape.

A diverse selection of naming controversies illustrates changing attitudes toward what's offensive as well as bureaucratic obstacles to wholesale revision. In 1963 Washington bureaucrats, embarrassed by toponyms containing *nigger,* ordered federal mapmakers to replace the N-word with *Negro,* then quite acceptable. Eleven years later a similar edict replaced *Jap* with *Japanese.* No other blanket changes followed, despite growing sensitivity among American Indians to toponyms containing *squaw,* an offensive reference that has evolved to connote whore in many indigenous languages. Lacking a universal substitute—alternatives include "Indian Woman," "Indian Girl," and "Clan Mother"—the U.S. Board on Geographic Names must evaluate each proposal individually, usually with input from state-level or tribal names officials. Although many *squaw* toponyms have disappeared since the early 1990s, when native activists concluded the term was derogatory, proposed replacement names are occasionally rejected as duplicative, inappropriate for the feature in question, or unsuitable because an individual commemorated by the new name has not been dead the required five years. Because board policy precludes merely erasing derogatory labels on maps, finding an acceptable substitute is a prerequisite to replacing a toponym based on *Negro, Chink,* or another pejorative.

Racial and ethnic insults are not the only questionable labels lurking on topographic maps. Irreverent miners and ranchers abetted by tolerant mapmakers enriched the national gazetteer with names like Brassiere Hills (in Alaska), Mollys Nipple (one in Idaho and two in Utah, among others), Outhouse Draw (in Nevada), and Whorehouse Meadow (in Oregon). Citizens who appreciate a bit of ribaldry and anti-establishment rudeness have resisted attempts to substitute synonyms like Naughty Girl Meadow, which briefly replaced its raunchy counterpart in the late 1960s. More numerous are toponyms like Squaw Humper Dam (in South Dakota), Squaw Tit (one each in Arizona, New Mexico, and Nevada; three in California), and variations like Squaw Tits (in Arizona), Squaw Teat (in Montana and Wyoming), and Squaw Teat Butte (in South Dakota and Wyoming), all of which simultaneously insult Native Americans and denigrate women. Additional instances survive only as "variant names," disallowed for use on government maps but still part of the historic record preserved electronically in the U.S. Geological Survey's Geographic Names Information System.

To grasp the magnitude of offensive toponyms and explore their geography, I mapped state-to-state variations for common pejoratives found throughout the country but most numerous in Rocky Mountain and Pacific Coast states. In addition to suggesting underlying cultural influences, these geographic summaries afford a dramatic overview of the eighteen different feature types once labeled *Jap* and the numerical dominance of *squaw* names, which greatly outnumber past and current instances of *nigger* and *Negro*. Even more dramatic are the facsimile excerpts of maps with contentious names like Dago Gulch (in Montana) and Nigger Pond (in southern New York)—explicit examples that should help readers understand the surprise, revulsion, or knowing resentment nonwhite Americans feel when shown ethnic insults once condoned by government mapmakers.

Facsimiles are equally effective in documenting cartographers' efforts to replace offensive toponyms. Particularly intriguing is an example from upstate New York, where Niggerhead Point became Negrohead Point on federal maps and more recently Graves Point, but only on official New York State maps. With its own statewide series of large-scale base maps and no obligation to conform to the federal board's decisions, New York seized the initiative in removing a lesser N-word subject to tedious piecemeal replacement at the federal level. Just as fascinating are before-and-after excerpts from topographic maps of Hawaii that confirm the restoration of diacritical marks required for the accurate rendering of names like Pu'u 'Ō'o, butchered as Puu Oo on previous editions. Native Hawaiians who value their linguistic heritage welcome the restoration of an indigenous orthography that mapmakers obsessed with standardization once rejected as too complex.

Additional facsimiles illustrate a more heavy-handed use of geographic names to claim territory, signify conquest, and discourage the return of refugees. Although this book focuses on the United States and Canada, one chapter looks at the renaming of settlements in Northern Cyprus and the eradication of Arab toponyms in Israel, and another examines efforts by sovereign nations to control representation of their territory and adjacent international waters by government and commercial mapmakers in other countries. Treatment of applied toponymy's significant international component includes a glimpse of issues addressed by the United Nations Conference on the Standardization of Geographical Names, held every five years since 1967; the

UN Group of Experts on Geographical Names (UNGEGN), formed in 1972; and the Foreign Names Committee of the U.S. Board on Geographic Names, which oversees the diplomatically sensitive treatment of overseas place names on maps by the State Department, the CIA, and other federal agencies. A third chapter examines conflict between American and international agencies that claim the right to name features in Antarctica, under the sea, and on our moon, other planets and their satellites, and assorted asteroids. Post–World War II advances in remote sensing and seafloor charting triggered a land rush in commemorative naming as well as inevitable feuds over jurisdiction and explorers' rights. The chapter also looks fleetingly at streets, shopping centers, and housing developments, named without federal interference by municipal officials and private developers but subject to objections by emergency planners eager to avoid confusing duplication.

Applied toponymy's rich, dynamic complexity precludes a treatment that is globally definitive. By consciously focusing on name controversies in the United States with a few Canadian examples added for flavor, I've highlighted a small but important part of a multifaceted process that merits further treatment of personalities, rules, and impacts. My strategy of imparting insight through a coherent collection of revealing case studies is both a strength and weakness of this narrative, which offers the understanding of a map historian attuned to the diverse connections of science, society, and mapping technology but lacks the depth of an insider immersed in the workings of naming conventions and cartographic bureaucracy. Readers interested in different takes on applied toponymy can benefit greatly from Richard Randall's *Place Names: How They Define the World—and More* (2001), Alan Rayburn's *Naming Canada: Stories about Canadian Place Names* (2001), and Naftali Kadmon's *Toponymy: The Lore, Laws and Language of Geographical Names* (2000). Randall, a retired Defense Department geographer, was executive secretary of the U.S. Board on Geographic Names for twenty years; Rayburn, who was executive secretary for many years of what's now the Geographical Names Board of Canada, wrote numerous columns on place names for *Canadian Geographic;* and Kadmon is a retired Israeli professor who served as chief cartographer of the Survey of Israel and was actively involved in numerous UN initiatives on geographical names.

As these authors point out, applied toponymy has important roles outside mapmaking insofar as place and feature names appear in diverse documents, including scientific reports, news stories, statistical compilations, leases and property descriptions, and various government directives, civilian and military. Whether applied toponymy is a branch of cartography is a moot point: if you take a broad view of cartography, as I do, it clearly is. If you consider cartography as merely map design and production, it obviously isn't. From a user's point of view, I think I'm right. Although the toponymist generally has more in common with the historian and historical geographer than with the land surveyor or GIS expert, decisions about geographic names reach more of us through maps than through any other form of communication. In much the same way that geographic names are embedded in most maps, applied toponymy is deeply rooted in mapmaking.

Although authors must accept responsibility for errors of fact, evolving rules and new decisions on the names of individual features inject an unavoidable obsolescence more obvious here than with my previous books. With parts of the story unfolding as I wrote, research was a bit like trying to hit a moving target. For example, while chapters 2, 3, and 4 rely heavily on a copy of the massive GNIS database I obtained in May 2003, some examples cited reflect later queries to the online version. And rather than tinker with earlier chapters to reflect proposed editorial changes to official guidelines and ongoing alterations of the National Map's names layer, I've chosen the sane expedient of conceding minor temporal inconsistencies and noting that my summary maps, statistical tabulations, and related citations are snapshots, slightly but not outrageously out-of-date by the time readers see them.

A final caveat: because this book is intended for general readers, academic geographers will notice little reference here to poststructural critical theorists, who've said much about maps in recent years but little about toponyms. Simply put, I've not found their work particularly useful, especially when tedious regurgitation of Foucault crowds out case studies and fosters gratuitous assumptions about power and impact.

*   *   *

Numerous people helped me appreciate the richness of geographic names and the intricacies of their standardization. Early in the project Ronald Grim, formerly with the Library of Congress and now at the Boston Public Library, generously shared his experiences on the U.S.

Board on Geographic Names. Roger Payne, the board's executive secretary and head of the U.S. Geological Survey's Geographic Names Office, contributed invaluable insights on applied toponymy and assisted with contacts at other agencies. In addition to providing priceless information about controversial names and federal policy, Roger made valuable comments on a draft of the manuscript, as did George Demko, formerly the chief geographer at the State Department and more recently on the geography faculty at Dartmouth College. At the USGS Geographic Names Office, Jennifer Runyon, Robin Worcester, and Julia Pastore were particularly helpful. I also appreciate the assistance of Chick Fagan, at the National Park Service, and Trent Palmer, at the National Geospatial-Intelligence Agency.

At Syracuse University, David Call, Karen Culcasi, and Pete Yurkosky helped compile an initial bibliography; Sue Lewis, of the Maxwell School Information and Computing Technology Group, got me up to speed with Microsoft Access; John Western, my colleague in the geography department, backstopped my rusty French vocabulary; Brian Von Knoblauch, in the department's Geographic Information and Analysis Laboratory, interceded when the network was cranky; and Joe Stoll, our department staff cartographer, contributed timely insights on Freehand, Photoshop, and electronic imaging. At more rarified levels, the Maxwell School and the Vice Chancellor's Office supported travel and the purchase of research materials.

In working out the details of a publishing contract, I drew on the wisdom of Anita Fore, Penny Kaiserlian, Bill Strachan, and Peter Webber. At the University of Chicago Press, my editor, Christie Henry, was enthusiastic and supportive throughout, and my manuscript editor, Erin DeWitt, proved that having a sense of humor is not inconsistent with knowing the *Chicago Manual of Style* like exegetes know the Bible. I look forward to working again with Erin Hogan and Stephanie Hlywak, who vigorously promoted my last few books. Matt Avery helped wrap the package with an elegant design. And at Waldorf Parkway, I have Marge and our five cats.

CHAPTER ONE

Naming and Mapping

Someday soon map collectors will discover cartographic insults. What I'm talking about are government topographic maps with features named Nigger Lake, Chinks Peak, or Squaw River—linguistic left-overs from less sensitive times, when mapmakers conscientiously but uncritically recorded local usage. Their inherent appeal to collectors is twofold: growing scarcity and intriguing anecdotes. Never a huge share of the national cartographic heritage, map sheets with offensive names become harder to find as government mapmakers, embar-rassed by their unwitting role in the social construction of race and ethnicity, rename features and issue new editions.

Ask how these names got onto the map in the first place, and you might find revealing glimpses of the area's past and some pleasant surprises. Take for instance, Niggerhead Point, a small, otherwise un-remarkable cape that shows up on older maps of the Lake Ontario shoreline in Wayne County, New York (fig. 1.1). Although the name sounds repugnant, it refers a bit backhandedly to a station on the Underground Railroad, an abolitionist enterprise that helped thou-sands of slaves escape to Canada.[1] Most pejorative place names have less noble origins, and as numerous examples illustrate, the search for a suitable replacement name can be surprisingly contentious.

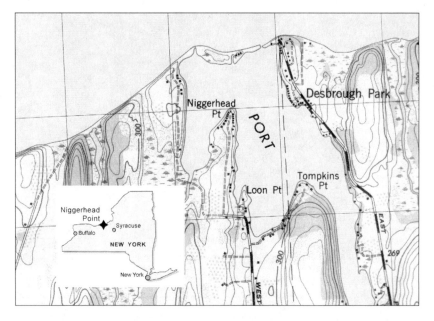

Figure 1.1. Niggerhead Point, as shown on the North Wolcott, New York, 7.5-minute quadrangle map, mapped by the Tennessee Valley Authority for the U.S. Army Corps of Engineers and published in 1943 at 1:31,680. (A quadrangle is a four-sided area bounded by meridians and parallels. A 7.5-minute quadrangle map covers 7.5 minutes, or an eighth of a degree, of both latitude and longitude.)

A case in point is Squaw Peak, which rises 1,800 feet above Phoenix, Arizona. The American Indian Movement (AIM), which made pejorative geographic names a part of its political agenda in the 1990s, considers *squaw* offensive to all Native Americans and has lobbied lawmakers for more than a decade to change the name. According to AIM, the word originated among French traders as a corruption of Indian slang for vagina; a trader who wanted a woman for the night asked for a squaw.[2] White settlers applied the term recklessly to hundreds of mountains, rivers, lakes, canyons, and other topographic features throughout the American West. Although Hollywood sanitized the meaning of *squaw* for filmgoers—in a typical shoot-'em-up Western, the dumpy, silent squaw acts like little more than her husband's property—Native American activists regard the term as both obscene and racist.

AIM's understanding of *squaw's* origin differs from that of place-names experts, who trace it to an otherwise neutral Eastern Algonquin word for young woman.[3] Although the negative connotation of a prostitute or sex slave apparently evolved after the term passed into English in the seventeenth century, for many white Americans *squaw* was and still is just a neutrally convenient one-syllable word for Indian woman. That many (if not most) American Indians find the word objectionable seems more deeply rooted in the white majority's often-brutal treatment of indigenous North Americans.

Some renderings of *squaw* are particularly offensive to women. According to Phoenix folklore, a local topographic landmark was called Squaw Tit Peak in the mid-1800s. As the area grew, white Christian women complained, and local officials shortened the name to Squaw Peak. This account surfaces occasionally in news stories, but I can't find cartographic evidence to support it.[4] History buffs who perpetuate the myth might be confusing Squaw Peak near Phoenix with two other named peaks in Maricopa County (fig. 1.2): Squaw Tit (singular) and Squaw Tits (plural), separate features about thirteen miles apart, are in the Sand Tank Mountains, over fifty miles south-southwest of Squaw Peak and sufficiently far from town to escape the censure of church ladies and indignant feminists.

A key obstacle to changing a name like Squaw Peak is finding an acceptable substitute. Arizona, like most states, has a state board on geographic names, which takes complaints and evaluates recommendations from interest groups, private citizens, and local governments. The board looks at the merits of the proposed substitute as well as its compliance with federal guidelines. A state board can approve a change, but if the U.S. Board on Geographic Names doesn't buy it, the new name never appears on federal maps, which include the U.S. Geological Survey's large-scale topographic maps, used not only by scientists and hikers but also by companies that make atlases, guidebooks, indexed street maps, and tourist maps. If the USGS doesn't pick up the change, commercial mapmakers will probably ignore it as well.

Substitutes rejected by the Arizona board include Iron Mountain and Phoenix Peak. Indian activists favored Iron Mountain, believed to be the English translation of the original Pima Indian name for the landmark.[5] Because naming authorities typically insist upon a historical link or other logical connection between a feature and its name, Iron Mountain seemed a promising candidate until anthropologists pointed out that the Pimas' Iron Peak was four miles away.[6] Unable to establish a historical association, the board rejected the proposed sub-

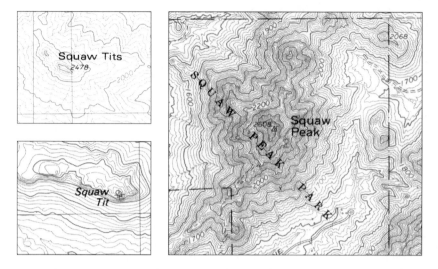

Figure 1.2. Squaw Tits and Squaw Tit, as shown on the Big Horn, Arizona (*above left*), and Kaka NW, Arizona (*below left*), 7.5-minute quadrangle maps, published at 1:24,000 by the U.S. Geological Survey in 1979 and 1986, respectively. Local mythology maintains that Squaw Tit Peak was an earlier name for Squaw Peak (*right*), as shown on the 1982 edition of the Sunnyside, Arizona, quadrangle map.

stitute. If the mountain were the site of a noteworthy magnetic anomaly or a nineteenth-century forge or smelter, Iron Mountain might have solved the problem. By contrast, the moniker Phoenix Peak, although locally connected, garnered no support. Why tag a prominent landmark with a locally unexceptional name?

Arizona's search for an alternative seemed hopelessly stalled until March 2003, during the Iraq War, when Lori Piestewa, a member of both the Hopi tribe and the U.S. Army's 507th Ordnance Maintenance Company, became the first Native American female soldier to die in combat.[7] Governor Janet Napolitano, who saw an opportunity to solve the Squaw Peak problem as well as honor a local war hero, called for renaming the mountain Piestewa Peak and aggressively lobbied the board to act immediately. On April 17, 2003, less than a month after PFC Piestewa's death, the board accepted the governor's proposal by a five-to-one vote.

Political expediency at work? You bet, but not without repercussions. The single "no" vote was a protest ballot, cast by the only board member who was not a state employee. Honoring a slain war hero might seem the obvious answer, but Governor Napolitano had coerced the board to waive its restriction, modeled on a federal rule against naming natural features after people dead less than five years. Another board member had resigned in protest, and the board chairman missed the meeting after the governor called for his resignation. State boards rarely act this precipitously, and remaining members smugly conceded that Arizona would need to wait five years for the federal names panel to consider the change.[8] A more immediate victory for the governor was the renaming of Squaw Peak Parkway, which the State Transportation Board had dubbed Lori Piestewa Freeway. The U.S. Board on Geographic Names has no jurisdiction over highways, shopping malls, and other man-made features. (Two notable exceptions are the names of canals [artificial rivers] and reservoirs [artificial lakes], which are subject to the federal board's policies.)

Washington's reluctance to dedicate geographic features to recently deceased persons is partly a reaction to its own hasty and coerced renaming of Cape Canaveral a week after the assassination of President John F. Kennedy in November 1963. Although commemorative renaming is a time-honored American way of coping with political assassination, the federal board might easily have resisted the frenzy of renaming that was sweeping the country that fall. What the board could not resist was intense pressure from Kennedy's successor, Lyndon B. Johnson, who insisted on renaming not only the NASA Launch Operations Center at Cape Canaveral, Florida, but also the cape itself. The president asked the Interior secretary to accelerate the review, and someone contacted board members individually by telephone.[9] All approved the change.

Florida residents were outraged. The government could call its missile-launch facility anything it liked, they reasoned, but the prominent projection eastward into the Atlantic would always be Cape Canaveral, a name derived from a Spanish term for reeds and rushes. One of the oldest mapped features in the state, Cape Canaveral had made its cartographic debut in 1564, on a map by French artist-explorer Jacques LeMoyne.[10] President Johnson had no right to change it, Floridians argued—phoning up the governor, to give him a heads-up, was hardly proper consultation. Pointedly the city of Cape Canaveral not only voted against renaming itself Cape Kennedy but also questioned the federal government's right to rename the cape. Since the slain leader was from Massachusetts, many asked, wouldn't it be more appropriate to rename Cape Cod? Florida's U.S. senators introduced a resolution calling for the restoration of the name Cape Canaveral and condemning the board for acting precipitously and without due deliberation, and thus illegally. Seen as disrespectful to the Kennedy family, the resolution died. In 1973, after Florida passed a law requiring Cape Canaveral on all state maps, the U.S. Board on Geographic Names accepted the state's position and restored the cape's original name (fig. 1.3). By general agreement, the NASA facility remained the Kennedy Space Flight Center.

Commemorative naming is one of ten ways geographic features get their names, according to George Stewart (1895–1980), a Berkeley professor of English whose books *Names on the Land* (1945), *American Place-Names* (1970), and *Names on the Globe* (1975) highlight the role of geographic names in the national cultural landscape.[11] Names can commemorate people, events, and even other places, as geographer Ren Vasiliev demonstrated in an insightful analysis of forty-eight places in the United States with *Moscow* in their name.[12] Historical research revealed that thirteen places or features were named after the Russian capital, the Moscow River, Moscow's church bells, or Napoleon's ill-fated attempt to take the city in 1812. Nine other American *Moscow*s are transfer names (also called shift names), relocated from one place to another. For example, when Moscow, Idaho, became large enough to need a name, the U.S. Post Office accepted the suggestion of Sam Neff, who ran the general store and was born near Moscow, Pennsylvania. Two *Moscow*s have anticipatory names (also called commendatory names), intended to confer prestige, and four more are mistake names, originally meant as something else. For example, the Moscow in Stevens County, Kansas, was initially Moscoso, named for an officer in Coronado's expedition; after residents shortened it to five letters, Mosco, a well-intended but meddlesome postal official added the *w* to make it look right. Twenty additional *Moscow*s remain in an "unknown" category because Vasiliev could find no clear explanation for their names.

Four of Stewart's other six categories are especially abundant on maps. Descriptive names like Muddy River and Dismal Swamp typically consist of two parts: a specific (like *Dismal*) followed by a generic (like *Swamp*), which usually identifies the type of feature. Although Muddy River is a good example, naming officials would probably reject it because most rivers, at least at flood stage, are indistinctively muddy. While this two-part specific-generic construction is also com-

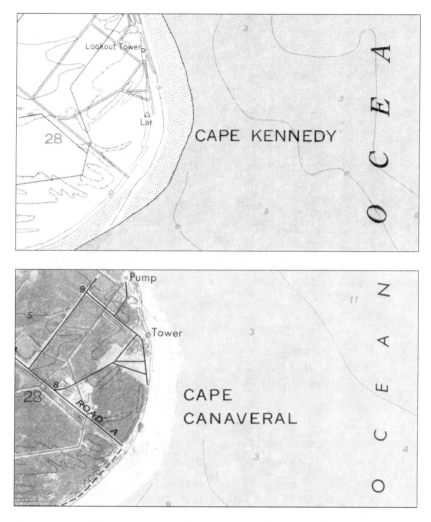

Figure 1.3. Excerpts from 1970 topographic (*upper*) and 1976 photomap (*lower*) editions of the U.S. Geological Survey's Cape Canaveral, Florida, 7.5-minute quadrangle map reflect the renaming of the cape in 1963 and the restoration of its original name a decade later.

mon among possessive names like Daggett Mills and Clems Ford (owned by or otherwise identified with Daggett and Clem, respectively), single-word renderings are widely used for populated places, as with Pottersville (for Potter's Town) or Pittsburgh. Specific-generic sequences are equally useful for associative naming, which identifies a specific feature by invoking its association with something else, perhaps a local landmark (as in Mill Creek), a plant (as in Fern Brook), or even another feature (as in Blue Mountain Lake). And incident names like Shipwreck Cove and Massacre River are similarly configured, with the specific referring to a noteworthy disaster, accident, lucky escape, or similar incident, and the generic identifying the type of feature.

Although less common, Stewart's two remaining categories are rich in intriguing anecdotes. Manufactured names include contrivances like Delmarva Peninsula (for the parts of Delaware, Maryland, and Virginia lying between the Atlantic Ocean and Chesapeake Bay) and Texarkana (a city overlapping both Texas and Arkansas). Housing developers sometimes manufacture street names by combining the first few letters of early residents' surnames, as in Yataruba Drive, a Baltimore address concocted to flatter the Yates, Taylor, Ruben, and Barton families.[13] By contrast, folk-etymology names typically evolve when a word that makes sense in one language acquires a wholly different meaning because of sonic similarity to a very different word in another language. In eastern Nicaragua, for example, the coast named for the Miskito Indians was reshaped in English as the Mosquito Coast, which invokes misleading images of insect-infested swamps. Although standing water there has its share of mosquitoes, they are no more numerous or vicious than elsewhere in Central America. As Stewart pointed out, some place names were consciously bestowed whereas others evolved through one or more intermediate versions.

Because Stewart focused on inhabited places rather than geographic features, his writings are largely silent on derogatory names, which rarely refer to regions or localities in which people live. For example, the pejorative Dago Gulch might survive on maps as an obscure feature name—there's one in western Montana (fig. 1.4), for instance—but Dagoville would never pass muster as an address. Even if postal officials agreed to list it, who would want to live there?

In examining controversial geographic names, I find it useful to distinguish between feature names and place names. Although Stewart maintained that "feature names" is merely shorthand for "place names of natural features," names like Squaw Peak and Dago Gulch have been far less subject to public scrutiny than names like Phoenix and Montana.[14] What's more, except in urban areas, named natural features are vastly more numerous than populated places, shopping centers, and other locations identified by name on large-scale topographic maps. When it's necessary to refer to both feature and place

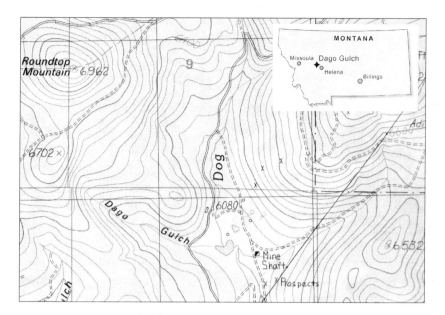

Figure 1.4. Dago Gulch, a feature name offensive to some Hispanic Americans, as shown on the Greenhorn Mountain, Montana, 7.5-minute quadrangle map published by the U.S. Geological Survey in 1989.

names collectively, the term *toponym,* from the Greek words for "place" and "name," is clear and convenient. Street names constitute a third category, addressed in chapter 8. Unquestionably geographic and certainly more numerous than feature names, they seldom appear on the U.S. Geological Survey's large-scale topographic maps. Rarely derogatory, street names can become highly controversial when local people disagree over commemorative renaming.

A few additional terms need definitions. Toponymy, which refers to the systematic study of the origin and history of toponyms, is part of onomastics, which studies all proper names. Toponymists have produced a substantial literature of place-name inventories focusing on a state, country, or region. Entries are organized alphabetically and typically include the place's location, by county or geographic coordinates, and a concise account of how and when it received its name. By contrast, applied toponymy is concerned largely with the standardization of geographic names.[15] In particular, mapmakers, sign makers, and news reporters need to know not only the current name for a place or feature but also its correct spelling. Naming authorities like the U.S.

Board on Geographic Names and its Arizona counterpart play an important regulatory role by ensuring that new names are meaningful and unambiguous.

As its name implies, the U.S. Board on Geographic Names is a panel of experts that meets regularly, reviews proposals, and sets federal policy for dealing with toponyms within and beyond the nation's borders. Although the board's decisions are most apparent on maps, it also governs usage in government scientific reports and other federal documents. Because foreign and domestic toponyms raise very different issues, the board divides most of its work between two key committees. The Foreign Names Committee, with representatives from the Central Intelligence Agency and the departments of Defense and State, deals with diverse concerns such as the transliteration of names rendered in non-Roman alphabets, the need for standardization within NATO and among America's military allies, and the dynamic renaming that accompanies marked political upheavals like the breakup of the former Soviet Union. By contrast, the Domestic Names Committee—with representatives from the Library of Congress, the Government Printing Office, the Postal Service, and the departments of Agriculture, Commerce, and the Interior—meets monthly, mostly to evaluate proposals submitted by state boards, federal agencies, and individuals. A recent addition is an online application form on the board's Web site, which encourages citizens to "propose a geographic feature name, suggest a name change, or indicate that a name is incorrectly applied."[16] In a typical year, the board receives between three and four hundred proposals, nearly evenly divided between new and substitute names.[17]

To guide its own deliberations as well as edify anyone eager to place a name on the national map, the board developed a set of principles, policies, and procedures. Four of the five principles declare a preference for the Roman alphabet and local usage, call for just one name and spelling for each place or feature, and acknowledge the traditional role of certain federal agencies in naming entities under their jurisdiction. For instance, the National Park Service names national parks and national monuments, whereas the Forest Service names the national forests. Should another Cape Canaveral case arise, Principle III (Names Established by Act of Congress or Executive Order) affirms the right of the president (or the Congress) to rename natural features without board approval.

A sixth principle, still under review as I write, indicates that no name can be removed from the official list. If a name is changed, the old name becomes a variant. A name can also be rendered historical, which requires a parenthetical annotation as in "Fort Sumter (historical)."

In contrast to the board's various principles, which reflect legal constraints and general philosophy, its ten policies, adopted between 1981 and 1996, refer more specifically to issues likely to arise in reviewing proposals. Policy III (Commemorative Names) is especially relevant to Arizona's effort to rename Squaw Peak.[18] It precludes commemorating living persons, calls for a five-year waiting period (increased in 1995 from the one-year wait established in 1984), and requires that the person honored "should have had either some direct and long-term association with the feature or have made a significant contribution to the area or State in which it is located."[19] Should Lori Piestewa's link to the Phoenix mountain prove problematic, the board could fall back on its rule that "an individual with an outstanding national or international reputation will be considered even if the person was not directly associated with the geographic feature."[20]

Two policies specifically address derogatory names. Policy II (Name Changes) asserts a preference for current local usage "whenever possible" and includes "a firm policy prohibiting the inclusion of a word in an official geographic name considered by the Board to be derogatory to any racial, ethnic, gender, or religious group."[21] Intended to lessen the likelihood of future controversies, this provision assumes somewhat optimistically that the board can reliably predict public attitudes decades ahead. Another provision reflects a resistance to duplicate names, especially common names that might lead to confusion. Distaste for duplication led the federal board to ask Piscataquis County, Maine, to rethink its blanket renaming of twelve *squaw* features by merely changing *squaw* to *moose*—as if Maine had a shortage of *Moose Lakes* and *Moose Hills*.[22]

Policy V (Derogatory Names) not only bars the adoption of any name the board finds "derogatory to a particular racial or ethnic group, gender, or religious group" but also encourages requests to change current names that citizens deem "derogatory or patently offensive."[23] Requesters must state why they consider the name objectionable and propose an alternative that satisfies the board's various guidelines. This policy obligates the board to consult with its state counterpart in finding an acceptable substitute and to "give careful consideration to all relevant factors, including the extent and distribution of usage, historical context, user perceptions and intent, and lexical meanings."[24] Occasionally a slight tweaking removes the sting, as in 1990, when Chinaman Spring, a landmark in Yellowstone Park named for a Chinese man who ran a laundry there, became Chinese Spring.[25] Most offensive names demand more drastic substitutes, such as Brunswick East for Jewtown, Georgia, and Corner Lake for Polack Lake, Michigan.[26]

A uniquely pointed sentence in Policy V warns that "the Board will not adopt a name proposal that includes the word 'Jap' or the word 'Nigger' whether or not it is in current local usage and regardless of by whom proposed." Although other provisions would surely rule out these two highly offensive pejoratives, the sentence refers obliquely to the only two instances in which the U.S. Board on Geographic Names ordered a blanket substitution for a derogatory name. In 1963, acting at the recommendation of Secretary of the Interior Stewart Udall, the board replaced all cartographic occurrences of *nigger* with *Negro,* and in 1974 a similar edict changed *Jap* to *Japanese.*[27] These were no doubt easy decisions insofar as *nigger* and *Jap* had become embarrassing as well as disgusting, and *Negro* and *Japanese* were obvious replacements. In the early 1960s, *Negro* had not yet acquired the distaste that led to its sequential replacement among more ethnically sensitive speakers, if not on maps, by *black, African American,* and *people of color.*[28] If *squaw* were more widely resented and had an obvious one-word stand-in, state names boards might not face the challenge of finding an appropriate substitute for each change.[29] As a further complication, tribal councils eager to preserve their cultural heritage often prefer a name from their own language.

How these pejoratives got on the map in the first place is largely a reflection of the mapmaker's role of recording names on the land, not censoring them. Most of the blame thus falls on local people who coined or perpetuated what the topographer merely recorded. Even so, by uncritically encapsulating local usage in a public document, mapmakers and the federal officials who oversaw their work made their successors responsible for defending or cleansing a cultural landscape tainted with ethnic and racial bias.

Ironically perhaps, in the early years of federal topographic mapping, when instructions about naming were vague at best, topogra-

phers had comparatively few opportunities to add pejorative names because their maps contained few named features. According to a report
published in 1904, two decades after the U.S. Geological Survey produced its first topographic map, federal topographers were working at
three scales: 1:250,000 (approximately four miles on the ground to an
inch on the map) for Alaska and "rougher portions of the Far West,"
1:125,000 (about two miles on the ground to an inch on the map) for
most of the remaining territory, and 1:62,500 (essentially a mile on the
ground to an inch on the map) for "more densely populated districts—
the Atlantic coast, parts of the Central region, and portions of the Pacific coast."[30] During the topographic survey's first decade, the USGS
preferred to map at 1:125,000 rather than 1:62,500 because the former
scale required only a quarter as many map sheets, but by 1893 coverage at 1:62,500 accounted for more than half the total area mapped
annually.[31] By contrast, the standard scale since World War II was
1:24,000 (exactly 2,000 feet on the ground to an inch on the map),
which accommodates far more named features than 1:62,500.

In addition to being comparatively expensive and time-consuming,
1:62,500 mapping was more difficult than its 1:125,000 counterpart to
keep current, especially for culture, a cartographic term for "features
constructed by man [such] as cities, roads, villages, and the names and
boundaries which are printed in black."[32] According to the Geological
Survey's *Manual of Topographic Methods*, published in 1893, for maps
at this more detailed scale "to be of value, [they] must undergo revision
at frequent intervals, in order to incorporate any changes in culture
and possibly in natural features due to natural or artificial agencies."
The *Manual* cautioned topographers "to limit the map to the representation of all natural features which are of sufficient magnitude to
warrant representation upon the scale, and to confine the cultural features, that is, the artificial ones, to those which are of general or public importance, leaving out those which are private in their nature."[33]

The 1904 report said nothing about toponyms, the recording of
which apparently was a routine and unremarkable part of cartographic
fieldwork, carried out during the warmer months and described in
considerable detail. A field party of surveyors measured distances,
angles, and elevations under the supervision of a topographer, who
would "walk or ride over all roads and paths and about the margins of
lakes, across country, etc., selecting routes so near one another that he
may be able to see all portions of the land."[34] Readers must infer that

relevant feature names somehow found their way onto the topographer's sketch maps before he returned to the office in the fall. The 300-page *Manual of Topographic Methods,* written "primarily for the information of the men engaged upon this work," was equally silent on mapping names.[35] It said nothing about polling local people or verifying spelling, but small examples of mapped topography accompanying a discussion of landforms confirmed the practice of naming peaks and other distinct features (fig. 1.5).

By 1928, when the Geological Survey issued a 432-page bulletin of *Topographic Instructions,* 1:62,500 was clearly established as the standard scale and applied toponymy was part of the process. "The importance of a complete and authentic record of feature names is so great," the new handbook asserted, "that nothing should be left for memory."[36] In particular, "all necessary notes should be made and all names recorded as soon as they are obtained." Tracings known as "names sheets" were registered to the upper and lower halves of each quadrangle map, so that mapmakers could record in place "all names

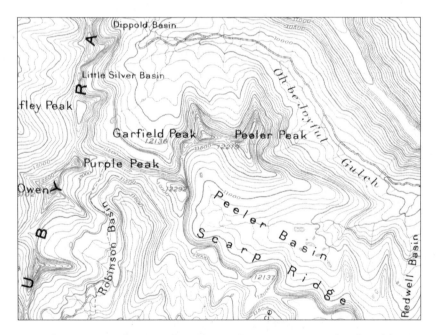

Figure 1.5. A sample topographic map in an 1893 instruction manual confirmed the need for feature names on the U.S. Geological Survey's early 1:62,500 topographic maps. Excerpt from Gannett, *Manual of Topographic Methods,* plate 18.

thought to be appropriate for the final map." A detailed list of "names to be shown" underscored the significance of feature names by including "hydrographic features," "relief features," and "springs, wells, and tanks, especially in arid regions where these features are of vital importance." And in contrast to USGS practice in 1904, the 1928 *Topographic Instructions* called for omitting names only to avoid graphic conflict with map symbols and other labels. To promote readability, an engraver could shift a name or use one of eighty-seven authorized abbreviations. Although a name might be omitted when there was no suitable alternative, "a position must be found for the name of every important feature."[37]

USGS policy called for toponyms to reflect local usage, which the mapmaker was told to confirm by consulting "at least two independent authorities for each name and spelling." Even so, map editors were asked to check names and features by consulting published maps, gazetteers, and reports of expeditions.[38] When discrepancies arose, "a definite effort must be made to obtain pertinent facts, so that a just decision may be reached in the field or made the basis for further reference." Censorship apparently was not an option. If field investigation indicated conclusively that a feature was known locally as Chinaman Spring or Dago Gulch, the topographer dutifully recorded it on his names sheet.

When a significant feature had no local name, making one up was an accepted practice. "In unsettled or sparsely settled regions," the *Topographic Instructions* advised, "it may be found desirable to give names to larger land and water features as a means of reference." Wary of abuse, the USGS warned that "the selection of new names should not be a mere matter of whim but should be made with due consideration of their geographic value." According to an Executive Order issued in January 1906, any names not based on local usage had to be approved by the U.S. Board on Geographic Names, known during the late 1920s as the U.S. Geographic Board. Under these procedures, topographers had little leeway for concocting pejorative toponyms.

The *Topographic Instructions* listed a selection of five explicit principles that guided board decisions. Although names that suggest "peculiarities of topographic features . . . such as their form, vegetation, or animal life are generally acceptable," overly common words like *elk, cottonwood,* or *bald* were discouraged.[39] Names of living persons were similarly suppressed, except for individuals of "great eminence." Equally suspect were "long and clumsily constructed names" and specific names with two or more words. Also disliked was "the possessive form of names" (like Jack's Ridge) unless "the object is owned by the person whose name it bears." Moreover, a river or mountain range was supposed to have only one name throughout its entire length. In addition to approving all new names, the board was to be consulted whenever "recognized authorities" disagreed over a local name or its spelling.

The intriguing history of American applied toponymy includes a few notoriously unpopular sweeping decisions a year after President Benjamin Harrison created the Board on Geographic Names in 1890. Harrison acted at the behest of several government agencies, including the U.S. Geological Survey and the U.S. Coast and Geodetic Survey, which was responsible for mapping the nation's coastline, harbors, and coastal waterways. Troubled by inconsistencies in spelling, board members voted to replace *centre* with *center,* drop the *ugh* from names ending in *orough,* and shorten the suffix *burgh* to *burg.*[40] Overnight, Centreview (in Mississippi) became Centerview, Isleborough (in Maine) became Isleboro, and Pittsburgh (in Pennsylvania) lost its final *h* and a lot of civic pride. The city was chartered in 1816 as Pittsburg, but the Post Office Department added the extra letter sometime later. Although both spellings were used locally and the shorter version had been the official name, many Pittsburghers complained bitterly about the cost of reprinting stationery and repainting signs. Making the spelling consistent with Harris*burg,* they argued, was hardly a good reason for truncating the Iron City's moniker—although Harrisburg was the state capital, it was a smaller and economically less important place. Local officials protested that the board had exceeded its authority. The twenty-year crusade to restore the final *h* bore fruit in 1911, when the board reversed itself—but only for Pittsburgh.[41] In 1916 the board reaffirmed its blanket change of *centre, borough,* and *burgh* as well as its right to make exceptions for Pittsburgh and other places with an entrenched local usage.[42]

Few of the board's rules reflect the impact of names on maps as pointedly as its efforts to defend federal wilderness areas against rampant commemorative naming. Defined by law as areas "where the Earth and its community of life are untrammeled [and] man himself is a visitor who does not remain," wilderness areas are protected from activities that undermine their wild character.[43] Banning motor vehicles is one way to keep "man's work substantially unnoticeable," and restricting toponyms is another. In 1985 the board formalized its wilderness-areas policy of "not approv[ing] names for unnamed features, including names in local use but not published."[44] Exceptions are allowed if the agency in charge can demonstrate an "overriding need," but the board is adamant that overstated "needs" not become loopholes.[45] In the wilderness as well as in cities, toponyms leave massive footprints on the cultural landscape.

CHAPTER TWO

The Quest for a National Gazetteer

In less than a hundred years, applied toponymy witnessed a transition as radical as our ancient ancestors' shift from hunting and gathering to agriculture—a fitting analogy insofar as topographers no longer roam the countryside inquiring about the names of hills and streams. American mapmakers had fully abandoned their traditional approach to determining local usage by 1990, when the U.S. Geological Survey completed fieldwork for the last of the more than fifty-five thousand large-scale quadrangle maps covering the forty-eight conterminous states.[1] Although keeping this massive topographic database up-to-date compels cartographers to uncover names for new shopping centers and other man-made features added since the last update, new or replacement names for physical features cannot get into the system without approval by the U.S. Board on Geographic Names. And because toponyms are potentially controversial, the board's Domestic Names Committee has a unique role in the politically wary federal mapping community. This chapter explores the technical and procedural details needed to understand the committee's response to offensive geographic names.

Advances in printing and data processing led to pronounced changes in mapmakers' treatment of geographic names. At the dawn of the twentieth century, a skilled engraver painstakingly copied the topographer's carefully inked manuscript map onto polished copper printing plates. Copper provided an easily worked surface for incising lines, letters, and other map symbols with sharp tools called burins or gravers.[2] Considerable skill and experience were required because names and other map labels had to be engraved backward, letter by letter, after scoring the surface with faint, easily removed guidelines. The malleable metal was ideal for removing flaws and guidelines, correcting mistakes, and updating map content because the engraver had only to rub the surface smooth with a burnisher before re-engraving the affected area. For larger alterations, he placed the plate facedown on a small anvil and hammered out the obsolete cuts from the back.

Well suited to smaller-scale maps with relatively few toponyms, copperplate engraving could not keep pace with the increased production of comparatively detailed maps compiled from aerial photography, starting in the 1930s. Clearly obsolete by the early 1940s, copper engraving disappeared completely by the mid-1950s. Nowadays a computer-controlled laser beam engraves the map electronically by adding names, boundary lines, and other map symbols to a film negative, from which the map image is transferred photographically to the printing plate.[3] If a name must be changed or added, the mapmaker need only generate a new film negative. And Web maps and electronic atlases bypass plates and printing presses altogether.

These snippets of printing history are relevant to applied toponymy because hard-copy topographic maps depend on color printing, which requires separate printing plates for each ink. In the 1880s the Geological Survey got by with only three colors: brown for contour lines and other relief features; blue for rivers, streams, lakes, coastlines, and similar hydrographic features; and black for boundaries, transport routes, toponyms, and other cultural features. Demand for greater detail led to the addition of green for woodland and red for certain roads and boundaries as well as the pinkish tint highlighting built-up areas where only landmark buildings are shown.[4] Adoption of a sixth color, purple, in the late 1960s reflected the increased importance of aerial photography in map revision as well as the need to control costs. On "photorevised" maps, which are not field-checked, all features and toponyms added since the previous field-checked edition appear in purple. Minimal doctoring of the other five "color separates" lets mapmakers expedite production and highlight recent changes.[5]

In much the same way that map printing forced the segregation of features and labels by ink color, advances in graphic arts technology encouraged mapmakers to group similar features into distinct layers called feature separates. Developed largely during the 1940s and 1950s, these changes were based on map-size sheets of plastic film.[6] Unlike paper, which shrinks or swells with changes in humidity, plastic film provides a dimensionally stable, geometrically reliable record of map features, in a format readily combined photographically into appropriate color separations. Registration pins and precisely punched holes assured exact alignment of symbols and labels when layers were merged. Feature separates proved especially efficient for comparatively volatile phenomena like roads and railways, which are more conveniently updated when grouped together on separate drawings.

Plastic scribing, as it was called, involved a variety of materials and techniques. The most distinctive medium was scribecoat, named for its thin opaque coating that was easily scratched off with a hand-guided graver with a chisel-sharp point. Scribing is superior to free-hand pen-and-ink drawing because it yields uniform lines with an exact thickness determined by the scribing tool. What's more, map labels composed in crisp type at a print shop, transferred to cellophane, and applied as stick-up type to a corresponding lettering "flap" could be integrated photographically with linework for the feature category or color.[7] For each feature category or printing ink, all toponyms were grouped together in a separate layer.

By the late 1980s, when electronic printing and the geographic information system (GIS) made plastic scribing obsolete, mapmakers were accustomed to treating topographic details as layers of geographic information. This conceptual shift anticipated GIS file structures that assign topographic features to layers dealing with elevation, hydrography, boundaries, and transportation, among others. While geographic names are treated electronically as merely another attribute of cartographic features with names, applied toponymy had already laid claim to its own database, overseen by the Domestic Names Committee. All named features are linked to this centralized names layer, which facilitates standardization and makes it easy to identify controversial toponyms. This role was well established by 2003, when the National Research Council panel that reviewed USGS plans for cartographic modernization recognized geographic names as a distinct and crucial layer of the digital cartographic database known as the National Map.[8]

Toponyms warrant distinctive treatment for another reason: a comprehensive geographic names database is an obvious twenty-first-century format for the gazetteer, a basic geographic reference tool typically published as a bound list of place names organized alphabetically and accompanied by a short description of each place and its location. Although reference atlases often include a "gazetteer index" listing populated places and important physical features by name, type, and location, the scope and scale of an atlas usually constrain the range of entries. By contrast, a geographical dictionary, which needs no maps, has ample room for a pithy description of a place's population, economic significance, and historical importance. Although maps are not mandatory, they are often inserted for their informativeness and visual appeal. *Merriam-Webster's Geographical Dictionary,* for example, enlivens its otherwise tedious pages with 258 maps describing the internal details and regional setting of places like Hong Kong and Myanmar.[9]

A comprehensive national gazetteer can be enormous, especially for a country that is expanding and filling in. In an essay in *Names,* the journal of the American Name Society, University of Illinois map librarian Robert White underscored the impact of growth by comparing gazetteers published in 1843 and 1873.[10] The earlier compilation, titled *A Complete Descriptive and Statistical Gazetteer of the United States,* contained roughly 16,000 entries and ran to 752 pages.[11] Three decades later *The Centennial Gazetteer of the United States* crammed 36,000 entries into 1,016 pages.[12] Writing in 1970, White concluded that a comprehensive national gazetteer was "no longer financially feasible" because the number of places and features requiring entries had increased to 2.5 million.[13] Government publication seemed an obvious solution to the gazetteer's size and dubious profitability, but the slow pace of names standardization troubled White. If every name needed board consideration and approval, compilation might take a millennium. In his opinion, the United States had not produced a comprehensive gazetteer since 1884 and seemed destined to ignore a 1967 recommendation by the United Nations Committee on Names Standardization that every country publish its standardized names in a national gazetteer.[14]

In focusing on comprehensive national gazetteers that were complete or well under way, White overlooked an early initiative of the U.S. Board on Geographic Names, which two years after its formation in 1890 began compiling lists, by state, of toponyms on USGS topographic maps.[15] Between 1894 and 1906, Henry Gannett (1846–1914), chief geographer at the Geological Survey and chairman of the Board on Geographic Names, published gazetteers for twelve states, two territories, and Cuba, which was under American control in the years following the Spanish-American War.[16] Typical of the level of detail are two entries in the *Gazetteer of Colorado,* which describes Nigger Hill as a "station in Garfield County on the Denver and Rio Grande Railroad" and Squaw Gulch as a "gulch in Teller County, tributary to Cripple Creek."[17] Each entry also mentions the name of the "atlas sheet" (quadrangle map) on which the feature appears.

Gannett's gazetteers were authorized and published by his employer, the Geological Survey, not the Board on Geographic Names, which he chaired from 1894 to 1914.[18] When the Geological Survey refocused its mission around 1906, the series of gazetteers ended.[19] A similar policy exists today: the board establishes federal policy on names standardization and renders decisions on questionable toponyms but has no funds or authority to compile databanks or print gazetteers. Nowadays, though, a portion of the annual USGS appropriation is earmarked for maintaining the geographic names layer and supporting board operations. The board publishes its rulings periodically, in decision lists that cover only a tiny fraction of the nation's toponyms, typically only newly proposed names and those known to be controversial.[20]

Even though gazetteers were not a priority, the federal board encouraged their development at the state level. Gannett had demonstrated the practicability of piecemeal, state-by-state compilation, which avoided the massive cost and daunting complexity of producing and publishing a single nationwide inventory. In addition to pacing expansion of 1:62,500 topographic coverage, state gazetteers would shift much of the burden to the states, which had more ready access to regional expertise and relevant historical materials. In its *Sixth Report,* published in 1933, the U.S. Geographic Board (as it was known at the time) "hoped when all the State gazetteers have been published, to gather these up in one volume which will serve as a National gazetteer." The report noted optimistically that "more than 30 States" had established "geographic boards, commissions, or similar bodies," several of which were working on gazetteers, "while others are accumulating material which awaits legislative help toward securing needed clerical assistance to put the material in manuscript form."[21] But with the Great Depression straining state budgets, few legislatures could support compilation even if Congress subsidized publication. The 95-page *Official Gazetteer of Rhode Island*, issued in 1932 by the U.S. Government Printing Office, was the first and only installment of the hoped-for series, which perished in a drastic reorganization of the board in 1934.[22]

Enthusiasm was still apparent in 1953, when a *Names* article by Lewis Heck summarized developments over the previous two decades.[23] Although Maryland was the only additional state to issue a complete names list, compilers in California, Missouri, Ohio, Oregon, South Dakota, and West Virginia had produced partial listings focused on the origins and history of place names, and similar studies were under way in Connecticut and a few other states. Heck called for the newly established American Name Society to strongly support development of state and local gazetteers, which could harvest the "previously unrecorded" toponyms picked up in the slow but steady expansion of comparatively detailed, 1:24,000-scale topographic mapping.[24] In addition to vastly increasing the number of feature names shown on 1:62,500 maps, the gradually evolving large-scale national map made it easy to inventory each name's location by geographic coordinates or quadrangle name.

A key obstacle to a comprehensive national gazetteer was the sheer volume of information involved. Heck guessed that single-line entries set in small type for the estimated 1 million names would require roughly six thousand double-column pages. Other obstacles were the slow pace of systematic large-scale mapping and the time and cost involved in not only compiling the initial listing but keeping it up-to-date. As chief of the Geographic Names Section at the U.S. Coast and Geodetic Survey, which produced detailed navigation charts for coastal waters, Heck was well aware of the complexities of map revision. Because the largely invisible seafloor was in continual flux and unmapped marine hazards could be deadly, coastal mapping was considerably more current than its topographic counterpart, especially for sparsely inhabited interior lands, where upgraded surveys and map revision were not pressing needs.[25] Even so, nationwide topographic coverage was inevitable, and because the public would eventually demand a complete, up-to-date, and conscientiously maintained list of the country's toponyms, the crucial policy question was not *if* but *when*. Optimistically Heck suggested that "perhaps the delay in starting a national gazetteer is not an unmixed evil, since the longer it is put off the more names there will be to put into it."[26]

A decade later Heck made an active contribution to the piecemeal, state-by-state effort. As lead author of *Delaware Place Names,* published by the Geological Survey in 1966 and prepared jointly by employees of the USGS and the Coast and Geodetic Survey in cooperation with the Board on Geographic Names, he helped develop a model 124-page list of "known geographic names which are now applied or have been applied to places and features" in the small, comparatively manageable, three-county state.[27] Entries included up to ten attributes: name, feature type, population or elevation, description, county, geographic coordinates, the year of a decision (if any) by the Board on Geographic Names, map number (keyed to an index map showing names of quadrangles), variant names (if any), and the name's history (if available from a reliable published source).

The Delaware gazetteer contains no *nigger* or *squaw* toponyms. The closest pejorative I found is an entry for Negro Island, hardly derogatory in the early 1960s.

> **Negro Island:** *slight elevation,* wooded, 0.5 mi. across, in marsh 0.7 mi. W of Primehook Beach and 6 mi. NE of Milton; Sussex County; 38°51′20″ N, 75°15′20″ W. (map 31)[28]

As with over half of the roughly three thousand entries, boldface type indicates a name found on "modern maps" and believed to "generally reflect present-day local usage."[29] According to an index map toward the front of the volume, map 31 is the Milton, Delaware, 7.5-minute quadrangle, mapped in 1943 and 1955. As far as I can tell, Negro Island made its cartographic debut on the 1955 USGS map, published at 1:24,000 (fig. 2.1). Although the feature might show up on earlier editions, it's missing from both the 1943 map, published at 1:24,000 in a civilian edition based on a hasty wartime survey directed by the Army Corps of Engineers, and a smaller-scale 1918 USGS map of the Cedar Creek, Delaware, 15-minute quadrangle, published at 1:62,500.[30] As figure 2.1 shows, a slight rise in the middle of a swamp was too

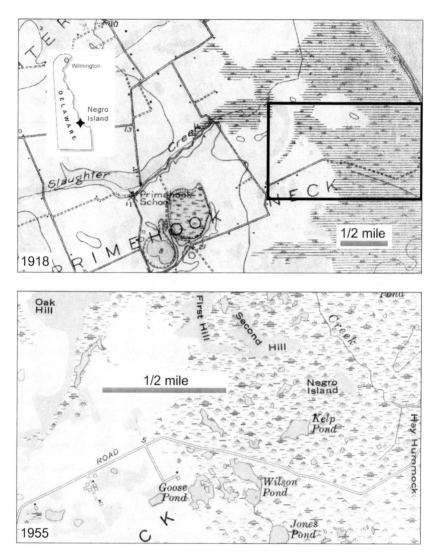

Figure 2.1. Comparison of 1918 and 1955 U.S. Geological Survey maps illustrates the omission of Negro Island, Delaware, from the earlier map. Upper facsimile is from the U.S. Geological Survey, Cedar Creek, Delaware, 15-minute quadrangle map, published in 1918 at 1:62,500; the rectangle identifies the area covered by the lower facsimile, from the U.S. Geological Survey, Milton, Delaware, 7.5-minute quadrangle map, published in 1955 at 1:24,000.

inconsequential for the more generalized 1918 map, which had no room for the tiny feature. The label for Negro Island thus reflects the increased number of named features picked up in the postwar expansion of 1:24,000 mapping.

Twenty years later the Geological Survey issued another list of Delaware place names. Its title, *National Gazetteer of the United States of America—Delaware 1983*, confirms Heck's hunch that the country would eventually undertake an official national gazetteer. Like the 1963 compilation, the new state gazetteer was a cooperative venture with the Board on Geographic Names. According to the introduction—prepared by Donald Orth, executive secretary of the board's Domestic Names Committee and a USGS employee—the list of "about 3,100" toponyms was part of a new series intended as "a national standard for reference and research on geographic names and a base for other data systems."[31] Because the board planned periodic revisions, year of publication became part of the title. That the introduction ignored Heck's 1963 compilation, which Orth had coauthored, most likely reflects a standardized format with essentially similar introductions for each state gazetteer. For instance, the New Jersey compilation, published the same year, had a virtually identical introduction, modified principally to note its inclusion of "about 10,000" names.[32]

The New Jersey gazetteer is noteworthy in several ways. In addition to listing three *Negro* toponyms, it contains two *squaw* names as well as an unquestionable pejorative, Nigger Pond. Clearly identified as a "variant" or unofficial name, the entry for Nigger Pond referred users to Potake Pond, within the bounds of the Sloatsburg quadrangle. Unlike the 1963 Delaware gazetteer, which included up to ten attributes per entry, the 1980s gazetteers arranged entries as rows in a table with eight columns of data: feature name, feature class, status, county, geographic coordinates, source coordinates (for the headwaters of streams), elevation (for peaks, lakes, and centers of populated places), and the name of the 7.5-minute quadrangle map. Variants appear twice, once in a separate line under the feature's official name and again, alphabetically, as a cross-reference to the official name, repeated below the variant name after the word *See*. To underscore a variant's inferior status, additional details are listed only in the line for the official name. Although a few variants refer to historical places that no longer exist, most are obsolete or duplicate names, seldom used locally but helpful to historians and genealogists.

Potake Pond is a cartographic curiosity. Its four variants (Negro Pond, Nigger Pond, Portage Lake, and Potake Lake) suggest that local residents were uncertain about not only the feature's name but also its standing as a pond or lake. The table reports the official toponym's status as "BGN 1938," a reference to the year the Board on Geographic Names ruled on the name. The need for a ruling is apparent in figure 2.2, which juxtaposes excerpts from maps published in 1910, 1938, and 1955. Mapmakers apparently recognized the need for a board ruling while compiling a completely new edition of the Ramapo, New York–New Jersey, 15-minute quadrangle map, which the Geological Survey published in 1938 at 1:62,500. The feature's location was apparently

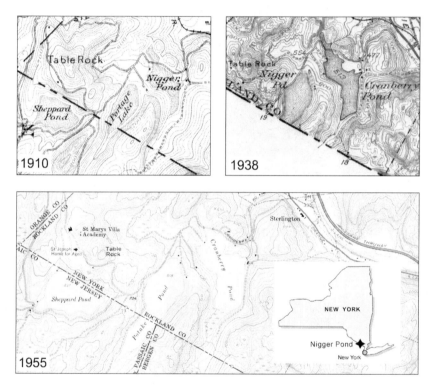

Figure 2.2. Excerpts from U.S. Geological Survey quadrangle maps for 1910, 1938, and 1955 record the curious history of the peripatetic Nigger Pond, a name applied at different times to features known in 1955 as Potake Pond and Cranberry Pond. Upper maps are from 1910 and 1938 editions of the Ramapo, New York–New Jersey, 15-minute quadrangle map, published at 1:62,500; lower map is an excerpt from the 1955 edition of the Sloatsburg, New York, 7.5-minute quadrangle map, published at 1:24,000 but reduced here to approximately 1:39,000.

more confusing than its name. Note that the 1910 map shows a Nigger Pond northeast of Portage Lake, partly in New Jersey. But on the 1938 map, what had been Nigger Pond is now Cranberry Pond, while the toponym Nigger Pond adorns the feature formerly known as Portage Lake, partly in New Jersey. Although *nigger* might have raised a few eyebrows, it's the wandering toponym that demanded board action. The *Decision List* for 1938–39 not only reports the board's ruling but describes the feature:

> **Potake Pond:** a small body of water crossed by the N.Y.-N.J. State boundary, about 4 miles northwest of Suffern, in Passaic County, N.J., and Rockland County, N.Y. (Not Portage Lake nor Nigger Pond.)[33]

Nowhere does the *Decision List* mention Negro Pond—perhaps a new variant created by the 1962 blanket replacement of *nigger* with *Negro*—but a companion ruling addressed the pond to the east and its contribution to the confusion.

> **Cranberry Pond:** a small body of water, 2 miles west of Ramapo, in the southwestern part of Rockland County, N.Y. (Not Nigger Pond nor Tivoli Pond.)[34]

Although board action deleted a toponym that was both derogatory and geographically erratic, the ruling had little immediate topographic impact. The 1938 map remained the area's most recent topographic portrait until 1955, when a 1:24,000 topographic map for the Sloatsburg, New York, 7.5-minute quadrangle wiped Nigger Pond off the cartographic landscape.[35]

Aware of the costly and painfully slow revision of its quadrangle maps, which are often twenty or more years out-of-date, the Geological Survey was no less concerned about the inherent obsolescence of its new state gazetteers. Intensive studies of several small areas suggested that the nation's large-scale maps and charts, however current and complete for their scale of publication, accounted for only 40 percent, roughly, of known feature names.[36] With many more toponyms to be harvested and an official mandate to do so, government officials were reluctant to continue the costly publication of printed state gazetteers, some as thick as a medium-size city's phone book. The USGS issued only six additional state-level compilations, for Kansas (1984), Arizona (1986), Indiana (1988), South Dakota (1989), North Dakota (1990), and Florida (1992), as well as a *United States Concise* volume (1990) with "a select list of about 42,000 geographic names . . . for major places, features, and areas within the United States and its territories."[37] Still committed to printed gazetteers, Don Orth introduced the *Concise* volume as "part of a series" that would "eventually list over three million names"—50 percent more than the rough estimate touted earlier in the decade.[38]

The decline of the paper gazetteer parallels the demise of paper as the principal medium for storing geographic information. Electronic technology, which transformed all aspects of cartography, proved enormously efficient for updating lists of toponyms. Computerized listings are readily revised, reliably reproduced, and easily managed once the database software is developed and debugged. But getting there was neither swift nor simple. Although the Board on Geographic Names began to view the national gazetteer as an electronic database in the 1960s, the Geological Survey, which maintains the data and provides computing support, lacked suitably efficient hardware and software until the mid-1980s.[39] As customized, up-to-date retrieval became more convenient, printed-and-bound state gazetteers gave way to CD-ROM gazetteers, on-screen viewing, and Internet downloads.

Ironically perhaps, the USGS names database was the key source for the 1.5 million place names in the eleven-volume *Omni Gazetteer of the United States of America,* published in 1991 by Omnigraphics, a Detroit firm that specializes in reference books.[40] Reference publishers survive by estimating market size and setting a price sufficiently high to recover production costs within a year or two. Research universities and large municipal libraries did not balk at the *Omni*'s $2,000 price tag, but few names aficionados bought their own set. Had the data not been readily available—government information is in the public domain, and thus freely copied and repackaged—the *Omni Gazetteer* would have been prohibitively costly even for Stanford and Yale. Its market was soon undermined by the Geological Survey, which in 1991 began selling inexpensive editions of more up-to-date names lists on CD-ROM and providing free copies to university reference collections and other libraries in the federal depository network. Despite competition from a government Web site with access to over 2 million names, the handsomely bound *Omni Gazetteer* looks impressive on library shelves and placates patrons (myself included) who appreciate the book format.[41]

Two key elements of the government's electronic national gazetteer are easily confused because of similar names and overlapping roles. Academic purists and federal bureaucrats differentiate the National Geographic Names Database (NGNDB), a collection of names files for every state and territory, from the Geographic Names Information System (GNIS) used to manage and access the data.[42] Both entities have acronyms, but the euphonious GNIS (pronounced "gee-nis," as for *genus*) is better known and more widely used. The surest way to keep the names and acronyms straight is to remember that the NGNDB is the largest database within GNIS, which also includes an electronic archive of the board's decision lists as well as lists of bibliographic references and the names of topographic quadrangles. Although I've simplified the discussion by treating GNIS as our electronic national gazetteer, GNIS software also supports compilation of the names database by Geological Survey mapmakers, the Board on Geographic Names, and the state names authorities that advise the federal board.

Circumstances dictated a two-stage work plan: Phase I to harvest feature names from the most recent, largest-scale USGS topographic maps and Phase II to incorporate toponyms from a variety of additional sources.[43] The first phase, initiated as a two-state pilot project in 1976 and completed nationwide in 1981, proved far easier than the second stage, which is still under way.[44] As of October 2003, when GNIS officials issued their latest status map (fig. 2.3), work was complete for forty-two states and well under way for another four. Because of budget constraints, Alaska, Kentucky, Michigan, and New York were still awaiting a systematic examination of gazetteers, local maps, county atlases, state records, historical documents, and other relevant sources. In tune with the federal government's preference for outsourcing, names compilation is contracted out, a state at a time, to university researchers or other qualified bidders.[45] Although the status map says nothing about the extensiveness of available source materials or the competence and commitment of contractors, a twenty-year range of completion dates suggests substantial unevenness in coverage, which is generally more accurate for states with a relatively recent completion date. Except for changes approved by the Board on Geographic Names and toponyms under the jurisdiction of the National Park Service or another federal agency, a state's gazetteer is essentially static once Phase II is complete. A third phase was contemplated, to ac-

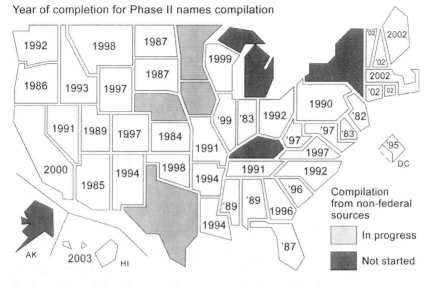

Figure 2.3. Status of Phase II geographic names compilation, as of October 2003. Redrawn from status map provided by U.S. Geological Survey, Office of Geographic Names.

commodate shifts in local usage, especially for areas with substantial growth, but this work will be carried out through partnership maintenance agreements with state and local governments and other federal agencies.

Phase II's slow, uneven pace also reflects the ambitious goal of picking up all nonstandard spellings and obsolete toponyms as well as the names of small, locally less important features omitted from 1:24,000 maps in order to "avoid map clutter" and the names of historical features no longer in existence.[46] Historically significant features that vanished from the cultural landscape include abandoned forts and Indian villages, landmark buildings razed for urban renewal, and train stations on abandoned rail lines; to avoid confusion with current features, GNIS appends "(historical)" in parentheses to these names. Equally important are toponyms for housing developments, subdivisions, shopping centers, highway rest areas, and other locally significant features deemed too ephemeral or inconsequential for 1:24,000 topographic mapping. Because GNIS is a "total information depository," researchers are advised to record "all obtainable names," except for streets, highways, and roads.[47] Instructions for Phase II compilation include a list of fourteen types of source documents that compilers should consult.[48]

Names lists maintained by various federal agencies also contribute to GNIS. For example, the Federal Aviation Administration provides names for airports, and the Federal Communications Commission identifies antenna sites for radio and television stations. Although these lists are merged into GNIS directly, other toponyms must first be recorded manually on the compiler's large-scale work maps, typically 7.5-minute quadrangle maps (now in electronic format), which also serve as a source for geographic coordinates and elevations. Annotations include an asterisk or a line marking the feature's position or extent, a bibliographic code identifying the source of its name, and "new" or "var" to distinguish a new name from a variant. A single set of conscientiously annotated work maps helps compilers assess their progress and avoid duplication. The maps are also useful in selecting an appropriate descriptor from an approved list of generic feature types like *range, ridge, summit*—in GNIS a mountain is never just a mountain. Although a degree in geography is not essential, compilers must understand feature types and know how to read and interpret topographic maps.[49]

Curious about Phase II's contribution to the study of derogatory toponyms, I searched the *Omni Gazetteer* and a mid-2003 snapshot of GNIS for names based on *nigger, niger,* or *nigar.* The printed gazetteer, based largely on Phase I data, yielded 111 features with an N-word pejorative, while the electronic repository turned up another 83, for a nationwide total of 194. State-level tallies were especially revealing: 31 states picked up at least one new insulting toponym, with the most spectacular gains in the West (fig. 2.4). California registered the largest increase, from 18 to 31, perhaps because of its size, population, and the relative newness of its Phase II names compilation, completed in 2000, while Nevada, with 11 new cartographic insults, was a close second. Stage 2 compilation no doubt explains most new entries, but it cannot account for the 13 post-*Omni* pejoratives in states with Phase II finished before 1990. No doubt some of these new N-word features reflect the work of GNIS officials, who enter updates and corrections based on their own research as well as information from state boards and the general public.[50]

All but three of the 194 offensive names are variants. Among this unholy trio, all are spelled with a single *g*, all lie east of the Mississippi,

Omni Gazetteer,
ca. 1990

One dot represents one feature with
nigger, niger, or *nigar* in its name.
All but three toponyms are variants.

No instances
in the state

Geographic Names
Information System
(GNIS), 2003

Figure 2.4. A state-level comparison of toponyms with *nigger, niger,* or *nigar* found in the *Omni Gazetteer* (1991) and a mid-2003 snapshot of GNIS suggest that Phase II names compilation has made a substantial contribution to the national gazetteer.

and all have a mildly curious history. The South's contribution is "Niger Post Office (historical)," a vanished feature in Escambia County, Alabama.[51] Because the name never appeared on a USGS topographic map, GNIS omits its latitude and longitude but assigns its location to the Roberts quadrangle. Nigers Creek, a stream in Geauga County, Ohio, lacks a quadrangle name as well as geographic coordinates, but GNIS dutifully reports a variant spelling, Niger's Creek, an apparent reflection of the federal board's aversion to apostrophes.[52] By contrast, the database confidently identifies Niger Hill, in Potter County, Pennsylvania, as a summit with an elevation of 2,360 feet, to be found on the Coudersport quadrangle at 41°51′40″N, 78°05′00″W.[53] I couldn't find it on 1969 and 1981 editions of the 1:24,000 map, which shows an unnamed 2,360-foot peak at this location, within a state-owned game

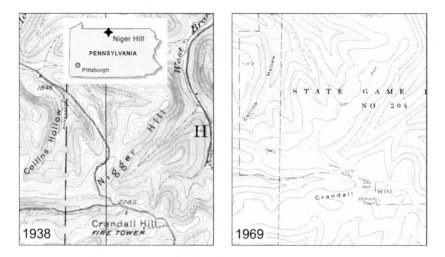

Figure 2.5. Niger Hill, a pejorative listed as a current toponym in GNIS through June 2003, reflects the erroneous entry of Nigger Hill, shown on the Coudersport, Pennsylvania, 15-minute map (*left*), published in 1938 at 1:62,500 but omitted from the Coudersport 7.5-minute map (*right*), published in 1969 at 1:24,000 (reduced here to 1:48,000). After a June 2003 correction, GNIS reports both spellings as variants of Negro Hill.

preserve (fig. 2.5, *right*). But the 1:62,500 USGS topographic map published in 1938 clearly identifies the feature as Nigger Hill—with two g's (fig. 2.5, *left*). My hunch is that when the Geological Survey mapped the area at 1:24,000 in the mid-1960s, state officials asked that the substitute name Negro Hill not be shown—an acceptable solution insofar as the feature is minor, and federal rules do not require names for all features. Although the single-g spelling probably got into the database as a keyboarding error, how it survived as a current toponym remains a mystery.[54] When I pointed it out to a GNIS administrator, he changed the feature's official name to Negro Hill and listed Nigger Hill and Niger Hill as variants. Who says academic research has no impact!

Like other aspects of cultural preservation, archiving obsolete toponyms in an electronic repository has advantages and disadvantages. On the plus side, GNIS cannot only enlighten historians, genealogists, and property surveyors confused by documents mentioning vanished places, but can also stimulate scholars fascinated by maps showing potentially meaningful patterns.[55] There's an intriguing doctoral dissertation, I'm sure, in the apparent concentration of *nigger* toponyms in the Rocky Mountain and Pacific states (fig. 2.4).[56] On the minus side, a historically complete names database combines an embarrassing reminder of our forebears' tolerance for racial insults with the lingering threat of an efficient mechanism for reapplying derogatory geographic names—not an altogether bad idea, though, should America ever achieve the racial equality needed to regard toponyms like Nigger Hill, Jap Valley, and Squaw Mountain as nothing more than quaint curiosities.

CHAPTER THREE

Purging Pejoratives

Derogatory names on government maps beg the question of what to do about them. Where there's a ready replacement, as in the substitution of *Negro* for *nigger* in the early 1960s, blanket change offers a reasonable solution—or at least it did until *Negro,* a formerly polite term that also connotes color, became offensive a decade later. Although *Negro* toponyms persist, many states now prefer a more locally or historically appropriate alternative, ideally the name (if known) of the specific African American for whom the feature was originally named. For *squaw* and other pejoratives without a ready substitute, blanket renaming is not an option, which means that state and tribal officials must find substitute names that satisfy the federal board's policies on commemorative naming, duplication, and historical and local relevance—a challenging task as examples cited earlier illustrate. In this chapter I explore more fully the conflicts and conundrums that arise in removing objectionable names from maps.

How extensive is the problem? To address this question, I compiled a list of derogatory terms and queried GNIS to see which ones might be found on topographic maps. The exercise was tentative insofar as names that are clearly derogatory in some settings can have plausibly innocent explanations elsewhere. For instance, a search for names based on *Chink* uncovered three official names and three variants.[1] Chink Creek, a short, wide stream just southeast of Baltimore, Maryland, might offend Chinese Americans who pull up its map (fig. 3.1), a couple mouse clicks away on the GNIS Web site, but the historical note for the Maryland stream reveals an innocuous root:

> A definitive name origin is unknown but it is believed the word "chink" is derived from the native term "chinqua" and translates to "large" or "great."

The electronic gazetteer turned up no instances of *kike* or *spic* as either official name or variant, but even if these words were to appear on a map, there's no guarantee they were ever intended to disparage Jewish people or Spanish speakers.

Because Native American names with a noble intent occasionally look or sound offensive in English, even long-accepted toponyms can raise eyebrows during a witch-hunt. In 1988, for instance, New York governor Mario Cuomo, spurred by a complaint about a wetlands map with a feature labeled Negro Marsh, called for state employees to wipe all derogatory names off state maps.[2] Eager to please the Executive Mansion, one industrious public servant proposed renaming Wappingers Falls, a village in the Hudson Valley about seven miles south of Poughkeepsie, because the first syllable, *Wap,* sounded suspiciously like *wop,* a slur on Italian Americans derived from *guappo,* Mediterranean slang for "thug."[3] New York also has a Wappinger Creek, Wappinger Lake, and the town of Wappinger, all named to commemorate the Wappinger Indians.[4] Although the Dutch wiped out most of the Wappinger in the seventeenth century, a few survived to fight for the

Figure 3.1. The GNIS Web site includes a link to TerraServer.com, which delivered this excerpt of the Dundalk, Maryland, 7.5-minute USGS quadrangle map showing the questionably derogatory Chink Creek.

American colonists during the Revolutionary War. As far as I can tell, neither Italian immigrants nor their descendants ever felt threatened by the name, and Wappingers Falls remains Wappingers Falls.

Among groups vulnerable to cartographic insult, Italian Americans get off lightly, perhaps because Italian immigrants mostly settled in eastern cities in the late nineteenth and early twentieth century, when naming was largely complete. GNIS turned up fifty-one official names and ten variants based on *guinea,* an anti-Italian pejorative that also denigrates people of mixed American Indian and African American ancestry. Most if not all of these toponyms refer, I'm confident, to guinea fowl or British coinage, and those that don't are probably too hazy in origin to be offensive. (Six of the ten variants were replaced by another *guinea* toponym.) As for names based on *wop,* I found only one, Wop Draw, a two-mile long valley in Wyoming about fifty miles north of Casper. I was unable to confirm an Italian connection because Wop Draw, like its neighbors Hunter Draw and Anderson Draw, was too inconsequential for Mae Urbanek's *Wyoming Place Names,* the latest word on the Cowboy State's toponyms.[5]

Another term offensive to Italian Americans is *dago,* a corruption of the personal name *Diego,* which also denigrates persons of Portuguese and Spanish descent.[6] GNIS uncovered twenty-one official instances of *dago* and nine variants. That seven of the variants reflect the substitution of *Italian* for *dago* suggests that Italian Americans were more likely to object, at least in Illinois, where this simple substitution helped rename Dago Hill, Dago Bowery (a subdivision in Chicago), and Dago Slough, and in New Mexico, where Dago Peak is now Italian Peak. Although *Italian* was used in renaming two Dago Creeks and a Dago Peak in Idaho, the state still has a Dago Creek, a Dago Peak, and a Dago Peak Gulch. By contrast, Montana's Dago Creek became Little Joe Creek and Nevada's Dago Pass became McKinney Pass. That most current *dago* toponyms occur in western states, where Hispanics are the dominant minority, suggests that Spanish speakers might be less sensitive than Italian Americans to this particular pejorative. Oddly, Minnesota, generally considered progressive and compassionate, still has a Dago Creek and a Dago Lake.

German and Polish Americans have even less cause for complaint. A query to GNIS found only eleven cases of *kraut* and six of *Polack.* Derived from *sauerkraut, kraut* is only mildly pejorative, and even humorous when we're not at war with the Kaiser or Der Führer.[7] Three instances (Kraut Run, Kraut Run Lake, and Kraut Run Lake Dam) are clustered around a single stream in Missouri, and two more (Kraut-man Creek in Missouri and Krautter Dam in Pennsylvania) might be commemorative. Only one *kraut* is a variant: Boone Creek, in western North Carolina, was once known as Kraut Creek. Its name might have been changed when waves of anti-German sentiment swept the country during the two world wars.[8] During the late 1910s Berlin, Iowa, changed its name to Lincoln, and East Germantown, Indiana, became Pershing, while in the 1940s Swastika, New Mexico, renamed itself Brilliant, and Germania, Washington, took the Indian name Wellpinit. Like *kraut, Polack* seems only mildly pejorative and is understandably less common on maps than *pollack*, a species of fish.[9] None of the *Polack* toponyms are variants, and some might be commemorative. Thanks to some not-so-clever turn-of-the-century immigration officers, Polack is now the surname of many Polish Americans.

Non-ethnic white Americans are not spared, at least not in the Southwest, where *gringo* has derogatory connotations. Derived from the Spanish word for Greek, *gringo* once referred to anyone speaking a strange language.[10] Now a disparaging reference to English speakers or North Americans, it makes rare appearances on Geological Survey maps. According to GNIS, Arizona has a Gringo Gulch, and New Mexico is home to Gringo Lake, Gringo Peak, and Gringo Water Well. All are official names, as is the toponym for Gringo, Pennsylvania, a tiny residential area about fifteen miles northwest of Pittsburgh. Although old maps indicate the name is at least a century old at this location, the hamlet was apparently too insignificant for *Pennsylvania Place Names*, a state gazetteer published in 1925.[11]

Dictionaries of slang and lists of objectionable terms circulating on the Internet turned up numerous words that are clearly derogatory when muttered in anger or scrawled spitefully in spray paint, but most of these pejoratives have other, wholly innocent meanings that make them ineffective as derogatory geographic names.[12] Although *spade* and *spook* are used to disparage African Americans, it seems more likely that shovels and ghosts inspired names like Spade Creek in Montana and Washington and Spook Hollow in Pennsylvania and Ohio. (Although a few of the thirty-three *spade* toponyms in GNIS might connote the suit of cards with the ominous ace or commemorate gamblers named Spade,[13] I doubt that any of the thirty *spook* toponyms refer to either spies or specific persons.[14]) And while the thousand-plus *coon* toponyms in GNIS are sufficiently common in the Southeast to have annoyed at least a few black people during the KKK era, these names most likely commemorate the raccoon, a medium-size mammal important in the diet of many backwoods southerners, black and white.[15] After all, if you want a geographic name that denigrates African Americans, why skimp with the ambiguous *coon* when the N-word itself was readily available? For similar reasons, I ignored toponyms based on *slant* and *slope,* which seem too ambiguous in a cartographic context to disparage Asian immigrants or their descendants.

Gook, as in Gook Creek, Michigan, is another matter. The word holds negative connotations among Vietnam-era veterans, who recognize the slur on North Vietnamese and other Asians. It's not a new pejorative either: during the Philippine-American War (1899–1902), when Filipinos resisted the U.S. occupation, *gook* filled a similar role, which it resumed a half century later during the Korean War.[16] Although the name's uniqueness on USGS maps—there are no other *gook*s in GNIS—suggests an indigenous origin, I didn't find it in either Walter Romig's *Michigan Place Names* or Virgil Vogel's *Indian Names in Michigan.*[17] Whatever its origin, *gook* both looks and sounds pejorative, which might explain its omission from the Chestonia, Michigan, 7.5-minute topographic map, published in 1983 at 1:25,000. The map labels other streams similar in length, and the creek's neighborhood in the woods of northern Michigan is hardly congested. That Gook Creek is clearly labeled on the smaller-scale Boyne City 15-minute map, published in 1961 at 1:62,500, raises the possibility of self-censorship at the Geological Survey. But careful inspection reveals a few slightly longer streams that also lack labels on the 1983 map. Nice hypothesis, though.

Another tactic in my search for offensively labeled features was to consult the catalog for a 1990 Library of Congress exhibit commemorating the hundredth anniversary of the Board on Geographic Names. Tucked away at the bottom of page 19 is a one-paragraph description of a display depicting "derogatory and suggestively derogatory names"—apparently the only mention of offensive toponyms in the exhibit.

Map. Color reproduction. "Gringo Peak," *Robinson Peak, New Mexico,* 1971. Color photographs: "Nigger Jack Hill," (changed to negro . . .),

Washington, California, 1951; "Jap Bay," (changed to Japanese . . .), *Kaguyak, Alaska*, 1954; "Dago Joe Spring," *Montezuma Peak, Nevada*, 1970; "Squaw Butte," *Verde Hot Springs, Arizona*, 1967; "Jewtown," *Brunswick East, Georgia*, 1956; "Chinks Point," *Annapolis, Maryland*, 1978; "Mick Run," *Spruce Knob, West Virginia*, 1970; "Polack Lake," *Corner Lake, Michigan*, 1958; "Scotchtown," *Dugger, Indiana*, 1963.[18]

A diverse list of ten toponyms, together with the names of their quadrangle maps, suggests that few ethnic groups or geographic regions escaped unscathed. The exhibit included the entire map for Gringo Peak, New Mexico, arguably the epithet affecting the largest number of Americans, but only photographically enlarged excerpts for the other nine. Parenthetical references to the two top-down blanket changes imply that federal cartographers had dealt with the most egregious examples, at least for non–Native Americans, and maps no older than the 1970s suggest misleadingly that offensive names were no longer appearing on newly reissued maps. Selection of five examples from states east of the Mississippi implies rightly that derogatory names are found in nearly all states but presents a geographically imbalanced portrait of a phenomenon that's notably more troublesome in the West.

Sorry if I appear to be picking on the Library of Congress. Its collections are outstanding, and staff at the Geography and Map Division have been enormously helpful in this and previous research projects. Still, I can't help feeling that whoever put together this collection of ten examples was stretching the notion of *derogatory* to meet a round-number quota. Simply put, Mick Run and Scotchtown are clearly not in the same league as Nigger Jack Hill, Jap Bay, and Squaw Butte. If they were intended as slurs on the Irish and the Scots, it's doubtful they had the intended impact. GNIS turned up only nine *Micks*, all of which might well commemorate persons (like Michael Philip Jagger) better known by their nickname.[19] And while the seven *Scotchtowns*, all official, sound like good places to buy tape or upscale whiskey, these plus 171 additional toponyms containing *Scotch* probably reflect more pride than prejudice.[20] In fairness, I must applaud the clever probity behind the catalog description's carefully worded phrase "derogatory and suggestively derogatory."

As an exemplar of offensive geographic names, Jewtown might seem more than merely suggestively derogatory, at least to anyone old enough to recall remarks about "Hymie" and "Jewtown" that got Jesse Jackson in trouble with the *New York Times* back in 1984.[21] Jackson was referring to a section of Chicago, around Maxwell Street, where he could buy clothes cheaper than in downtown stores. There's no Jewtown on USGS maps for Illinois, but there's one in coastal Georgia and another in central Pennsylvania (fig. 3.2). Both are small settlements easily ignored by Rand McNally and the AAA. Georgia's Jewtown, featured in the Library of Congress exhibit, was apparently named in the late nineteenth century for two Jewish brothers who ran a store there.[22] These days the African Americans who inhabit this historic waterfront community feel intense pressure from rising land values. If they sell out to condominium developers offering more than $100,000 an acre, Jewtown will most likely drop off the map.

Is *Jew* by itself inherently offensive? Sure is, according to the multiculturalist journalists who prepared a "Dictionary of Cautionary Words and Phrases."[23] "Some people find use of Jew alone offensive and prefer Jewish person," they warn. In a similar vein, the *American Heritage Dictionary* deplores "Jew lawyer" as "both offensive and vulgar" but thinks "Jewish lawyer" acceptable.[24] These strictures obviously predate the naming of Jew Point (in Florida), Jew Peak (in Montana), and the other fifteen similar toponyms in GNIS.[25] Probably commemorative, these toponyms sound at least mildly offensive, but perhaps not as much as *jewfish,* a tasty bottom-feeder that inspired the names of Jewfish Point, California, and nine features in Florida.[26] Although the name has raised eyebrows of vacationing northerners, it seemed secure until Gary Grossman, a University of Georgia ecologist, successfully lobbied the American Fisheries Society to rename the fish the goliath grouper, a fitting Old Testament moniker for a species that can grow to eight hundred pounds.[27] The society's decision, announced in 2001, prompted a few requests to change feature names, but the Board on Geographic Names wants appropriate substitutes as well as a nod from the state names board.[28] Lack of interest among local Jewish leaders, who don't see *jewfish* as an anti-Semitic slur, suggests that relabeling a handful of small features might prove a lot more difficult than renaming a big fish.[29]

Like their Jewish counterparts, Chinese American leaders seem to have bigger fish to fry than mildly offensive toponyms like Chinaman(s) Hat, which describes broad, cone-shaped summits in Idaho, Montana, Oregon, Texas, and Washington. Although the name carries

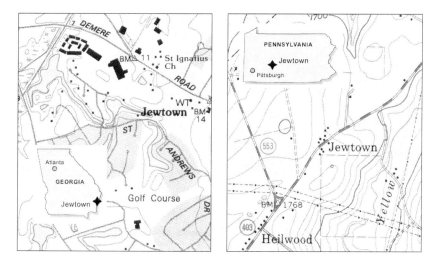

Figure 3.2. American cartography's two *Jewtowns*, as portrayed on U.S. Geological Survey 7.5-minute topographic maps. Excerpt on the left is from the Brunswick East, Georgia, quadrangle map, published in 1993 at 1:25,000, and excerpt on the right is from the Commodore, Pennsylvania, quadrangle map, published in 1993 at 1:24,000.

negative connotations of Chinese laborers brought in to build railways, only five of the thirty-five *Chinaman* or *Chinamans* toponyms in GNIS are variants that suggest replacement.[30] The instance that ruffled the most feathers is Chinaman Spring, a thermal spring in Yellowstone National Park that's too close to larger, more significant features, like Old Faithful, to show up on USGS topographic maps. It wasn't too small, though, to warrant a sign mixing a humorous anecdote with the negative stereotype of a Chinese hand-laundryman. The entry in Mae Urbanek's *Wyoming Place Names* tells the story:

> **Chinaman Thermal Spring,** YNP: normally dormant; in the early days of the park, an enterprising man put in soap, and proceeded to wash clothes in its warm waters; by chance it erupted 40 feet high, as it sometimes does, throwing clothes and soap skyward; hence the name.[31]

Chinaman apparently did not sit well with a Chinese American visitor from Hawaii, whose objections led the National Park Service to propose renaming the feature Chinese Spring.[32] Satisfied that the revised toponym was historically correct, the federal board approved the change in 1990.

This was not the only time signage highlighted a questionable toponym. In the early 1990s, a prominent marker on U.S. 40 in western Maryland led to complaints about the name of a linear ridge running northward through Maryland into Pennsylvania. It's a remote area, and the sign reporting Negro Mountain's elevation (3,075 feet) had little competition. The name commemorates the bravery of Nemesis, a slave killed in a 1774 skirmish with the Indians while fighting alongside his owner, Colonel Thomas Cresap.[33] Nemesis was buried on the mountain, named in his honor more than two centuries ago. Among the complaints was a formal proposal to rename the ridge Black Hero Mountain, which local historians resisted vigorously. According to Maryland's state archivist Edward Papenfuse, Negro Mountain is a historically significant name that "reflects an eighteenth century sensitivity to the important contribution African Americans made that is rarely so publicly demonstrated." Retired schoolteacher Marguerite Doleman, who set up a black history museum in her Hagerstown home, preferred the old name too. Had the ridge been named Nemesis Mountain, she told the *Washington Post,* the sign never would have "raised the question in my mind: What Negro?"[34] Convinced that the current name was both respectful and historically rooted, the Board on Geographic Names rejected the proposal.

Four decades ago the name Negro Mountain would not have been problematic. Newspapers used *Negro* in headlines and stories, the Census Bureau reported population statistics for Negroes, and most educated Americans, black and white, considered the Spanish word for black more civil than *colored* and definitely preferable to the N-word, lurking explosively in countless maps and gazetteers. In 1962, when the Board on Geographic Names responded to Secretary of the Interior Stewart Udall's complaint about geographic names containing *nigger,* substituting the polite and respectful *Negro* seemed both humane and logical.

Although *Negro* later lost favor with African Americans and the media, it never evoked the disgust apparent in Udall's July 27, 1962, letter to board chairman Edward Cliff.

Whatever the overtones of the word ["nigger"] were in the past, unquestionably a great many people now consider it derogatory or worse. It is like an obscenity in that avoidance of its use is common courtesy and in that its use may incur some sort of social penalty. I do not see how the Federal Government can in conscience require the use of the word in any connection.

I am aware of the need to discourage attempts by individuals to change perfectly good names for purely personal reasons. However, the attitude toward "nigger" is broader than personal, and a name can hardly be "perfectly good" if it contains a word that many people find offensive or will not say out loud.[35]

Cliff's response promised the secretary thorough if not prompt action.

All such names currently appearing on Federal maps will be reviewed by the publishing agencies, and as the maps are revised or reprinted, the names will be modified to remove any derogatory implications.[36]

A decade later the word *Jap* met Udall's "out-loud" test. On January 8, 1974, the federal board discussed the embarrassment of numerous *Jap* toponyms in western states, especially Alaska and Oregon (fig. 3.3). According to GNIS, these two Pacific Coast states each accounted for a half dozen of the nation's thirty *Jap* toponyms, all variants now, thanks to a board resolution.[37]

AGREED that the name "Jap" in geographic names be considered derogatory, and is to be avoided in Federal publications by changing the name to "Japanese," "Nisei," or by formally proposing another acceptable name.[38]

Although the board's resolution proposed both *Japanese* and *Nisei* as potential replacements, the latter must have sounded too foreign for federal cartographers: there are no *Nisei* toponyms in GNIS. I like to think that blanket erasure of anti-Japanese pejoratives reflects both the convenience of a ready substitute and board members' empathy for the more than 100,000 loyal Japanese Americans forcibly "evacuated" to detention camps during World War II.[39]

While Japanese Americans seemed content with the replacement of *Jap* with *Japanese,* African Americans seemed less sanguine about the blanket substitution of *Negro* for *nigger.* Curious about the acceptability of this universal substitute, I looked at features with a variant name containing the N-word and a new name formally endorsed by the federal board after 1962.[40] Of the 77 features so identified, only 24 still have *Negro* in their official name. (Why these 24 names warranted

Jap Creek
Jap Hills
Jap Gap
Jap Creek
Jap Creek
Jap Bay

Jap Gulch
Jap Lake
Jap
Springs
Jap Creek
Jap Hollow
Jap Meadow
Jap
Creek
Jap Creek
Jap Park
Jap Davidson Ditch
Jap Creek
Jap Mine
Jap Point
Jap Valley
Jap Ditch
Jap
Ranch
Jap Slough

ALABAMA
Jap Creek

ARKANSAS   MINNESOTA   TEXAS
Jap Knob   Jap Lake   Jap Smith Branch

FLORIDA   OKLAHOMA   WEST VIRGINIA
Jap Rock   Jap Beaver Lake Dam   Jap Post Office

Figure 3.3. The nation's thirty *Jap* toponyms once libeled a wide range of features, largely in western states. Thanks to the Board on Geographic Names, which substituted *Japanese* for *Jap* in 1974, all are now variants.

board action at all is a bit puzzling—perhaps the board was consulted to clarify blanket renaming, which was unusual.[41]) Of the 53 features with a post-1962 board decision date and an official name no longer containing *Negro*, 32 now have a toponym that looks commemorative. Examples include Harrison Canyon (in California), John Brown Lake (in Michigan), and Reub Long Butte (in Oregon). By contrast, the remaining 21 features have nonpersonal names like Wolf Lake (in Michigan), Lone Pine Lake (in South Carolina), and Freedom Lakes (in Wisconsin). Renamers evidently would rather commemorate ancestors, local heroes, and other specific persons than memorialize plants, animals, or abstractions.

Despite the emergence of *African American* as the preferred label for persons of African descent, none of the 53 renamed features has a new name containing *African* or *Africa*. Unlike their Japanese American, Chinese American, and Italian American counterparts, African Americans apparently have little interest in memorializing either their ancestral continent or places therein. For physical features and places other than cemeteries, churches, schools, and other buildings, GNIS lists 60 official toponyms containing *Japanese* or *Japan*, 459 based on *Chinese* or *China*, 109 with *Italian* or *Italy*, but only 26 with *African* or *Africa*.[42] In a more geographically refined search, I found no toponyms honoring Ghana, Kenya, or Nigeria, and only four for Liberia, which identifies populated places in North and South Carolina, a park in Texas, and a historic settlement neighborhood near the New Haven, Connecticut, waterfront.[43]

Why this reluctance to commemorate geographic roots? An answer might lie in the abortive attempt by black activists in East Palo Alto, California, to rename their community Nairobi. Younger residents of this predominantly African American San Francisco suburb generally supported renaming, while older residents strongly resisted the change—by a greater margin than white residents, according to a survey of voters following a 1968 public referendum.[44] Older black residents had moved to the area to escape the ghetto and wanted no part of a toponym with built-in bad memories. As one resident told a reporter for the *New York Times*, "With a name like Nairobi, everyone will know that we are black."[45]

If black pride ever mounts a massive effort to replace *Negro* toponyms, there's considerable work ahead for state officials as well as the federal board. According to a map of features with names containing *Negro* (fig. 3.4), most of the effort would occur in the southern half of the nation, in both the Southeast and the Southwest. To present a more realistic picture of the task, I omitted cemeteries, churches, schools, and buildings, for which names can change without government approval—included in GNIS as landmarks, cemeteries and structures are typically too small or otherwise insignificant to appear on topographic maps. I also omitted historical names, unlikely to change, as well as Spanish names such as Cerro Negro, which means "black hill."[46] Spanish toponyms are especially common in New Mexico, for which the map omits thirty instances of *Negro* that struck me as obviously Spanish. These omissions, admittedly conservative, are warranted because the Land of Enchantment state is no more likely to rename its Monte Negro (black mountain) than South Dakota is likely to rename the Black Hills. Should renaming begin in earnest, states in the upper Midwest will have a lighter burden than California, Tennessee, and Texas.

Although patchy renaming might reduce the number of *Negro*-based toponyms potentially offensive to African Americans, new in-

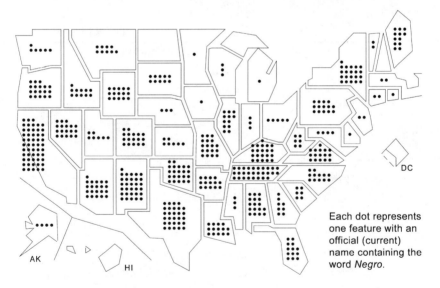

Each dot represents one feature with an official (current) name containing the word *Negro*.

Figure 3.4. Features with names containing *Negro*. The toponyms represented do not include cemeteries, churches, schools, and other buildings; historic names; and names that are more ostensibly Spanish in origin.

stances will surely arise. Phase II names compilation, which is still under way in several states and not even begun in others (see fig. 2.3), can add to the count, as can revelations of substantial Phase II omissions. In March 2003, for instance, two North Carolina legislators introduced a bill to ban derogatory place names on official maps, tax rolls, and deeds. Titled "Abrogation of Offensive Geographical Place-Names," the bill targeted seven specific toponyms.

SECTION 2.(a) The geographic place names set forth in this section are deemed to be offensive and insulting:

(1) Nigger Bay – Currituck County.
(2) Nigger Head – Clay and Madison Counties.
(3) Niggerhead Creek – Union County.
(4) Niggerskull Creek – Jackson County.
(5) Nigger Skull Mountain – Haywood County.
(6) Niggerskull Mountain – Jackson County.
(7) Nigger Spring – Haywood County.[47]

The bill's authors had apparently found them in William Powell's *North Carolina Gazetteer*, published by the University of North Carolina Press in 1968, five years after the Board on Geographic Names banished the N-word from federal maps.[48] Curiously, the compiler for North Carolina's Phase II, completed in 1992, had somehow overlooked numbers 2, 4, 5, 6, and 7, which were not listed in GNIS as variants. What's more, the sanitized versions, with *Negro* replacing *nigger*, were missing as well, probably because the features themselves never appeared on federal maps. Informed of the omissions by news reporters and state officials, the USGS Geographic Names Office promptly updated its database.[49]

As the North Carolina law indicates, geographic names have an important role in the legal descriptions of land parcels, where a name like Niggerskull Creek, hiding among the tedious details of survey traverses and subdivision plats, can pop up unexpectedly in real estate transactions or disputes over property taxes or rights-of-way. Offensive land titles are more likely in eastern states, where boundaries typically conform to terrain features like streams and ridges. In central and western states, by contrast, parcel descriptions commonly reflect a rectangular land-survey system based on artificial grid lines rather than natural features.[50] The relevance of old deeds to new disputes is a strong justification for preserving obsolete toponyms, however offensive, as variants.

In singling out names based on the N-word, North Carolina's new derogatory names statute pointedly ignores *Negro*-based toponyms, which most African Americas ignore or grudgingly tolerate. Persistence of these racial references on the cultural landscape reflects both the comparative mildness of *Negro* as well as the bureaucratic hurdles of applied toponomy. Red tape apparently stymied Gary Bledsoe, president of the Austin, Texas, chapter of the National Association for the Advancement of Colored People (NAACP). According to a 1990 article in the *New York Times,* Bledsoe objected to the name Negro Creek, upgraded from Nigger Creek by universal renaming three decades earlier. "Considered an improvement in 1962," the story noted, the name "sounds unacceptable in 1990."[51] The article noted the willingness of the Board on Geographic Names to adopt a suitable replacement if the Texas Office for Geographic Names Coordination agreed. Bledsoe seems to have abandoned his crusade: a LexisNexis search uncovered no follow-up stories, and the GNIS file for Texas contains only one board-approved change involving a *Negro*-based variant: in 1984, probably in response to a petition from Spanish speakers, Negro Creek became Arroyo Negro.

Federal approval is useful but not mandatory. As Arizona demonstrated in renaming Squaw Peak to honor a soldier killed in 2003 in the Iraq War, governors and legislators don't need permission from Washington to change toponyms on deeds, tax maps, and state tourist maps. But as figure 3.5 illustrates, it helps if a state has its own large-scale base maps, like the nearly one thousand 7.5-minute quadrangle maps maintained by the New York State Department of Transportation. The three map excerpts focus on Niggerhead Point, an upstate New York feature introduced in chapter 1 (fig. 1.1) and renamed Negrohead Point sometime between 1943 and 1955, well ahead of the blanket renaming of 1962. After *Negro* fell from favor, state mapmakers renamed it again, in 1977 or earlier, without the consent of the federal board. Although local fishing maps reflect the new name, Graves Point, it's still Negrohead Point on USGS maps and NOAA (National Oceanic and Atmospheric Administration) hydrographic charts, which rely on GNIS.

If Graves were the name of a locally prominent African American,

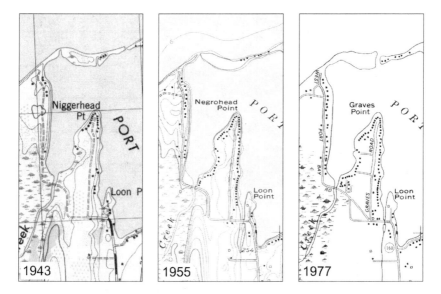

Figure 3.5. The cartographic transition of Niggerhead Point, on Port Bay in upstate New York, to Negrohead Point and more recently to Graves Point. Center and rightmost maps were reduced to 1:31,360, the scale of the leftmost map. Left and center maps are U.S. Geological Survey topographic maps published in 1943 and 1955, respectively, for the North Wolcott, New York, 7.5-minute quadrangle. Rightmost map is a planimetric quadrangle map published in 1977 by the New York State Department of Transportation. New York State recognizes the name Graves Point, but because the Board on Geographic Names never approved the toponym, federal maps still identify the feature as Negrohead Point.

rather than (I'm guessing) a reference to burial plots, official action by the federal board could give his or her accomplishments wider recognition in the "history notes" accompanying GNIS records like the entry for Connecticut's Walker Pond, in the city of Milford, near New Haven. Known as Nigs Pond until 1996, the feature is now "named for Reverend Charles D. Walker (?–1987), a prominent Milford resident and a founder of the Milford Council on Aging." Because the name Nigs had to go and no one wanted to revert to the earlier toponym, Negroes Pond, commemorating Walker was both appropriate and expedient.[52] If the respected reverend had died more than five years earlier, greater New Haven's substantial black population would surely have yielded another African American worth honoring—which suggests that the persistence of *Negro* toponyms in many remote rural locations might reflect a combination of inadequate historical records, a dearth of local notables, and an absence of African American groups with sufficient interest and savvy to initiate commemorative renaming.

But what of *squaw,* which accounts for 785 official toponyms, predominantly in the Pacific and Rocky Mountain states (fig. 3.6)? Although it has yet to rival *nigger* in "out-loud" offensiveness, *squaw* clearly upstages *Negro* as the thorniest issue in applied toponymy, thanks to Native American activists who put the S-word on their agenda in the early 1990s. It's an intriguing example of how a small number of people—victims and activists—can demonize a name to the point where group members and substantial numbers of outsiders feel the pain and demand action. From relative obscurity as a rude term for an American Indian woman or wife, *squaw* entered the new millennium as a hate word that connotes whore or vagina. Although many Americans reject the activists' etymology as yet another excess of political correctness, names authorities at state and federal levels see a huge yet unavoidable task of cartographic cleansing, especially in states like California and Oregon, with 104 and 166 *squaw* toponyms, respectively.

Are these the worst states? Not necessarily. Because bigger states have more nameable features than smaller states, it's necessary to adjust the counts for size. Because differences in both the intensity of naming and the thoroughness of Phase II names compilation might

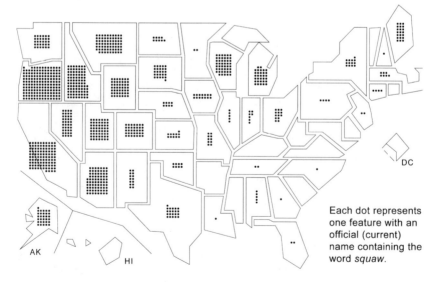

Each dot represents one feature with an official (current) name containing the word *squaw.*

Figure 3.6. Features with official names containing *squaw.*

distort the picture, dividing the number of *squaw* toponyms for each state by its total number of geographic names is more logical than dividing by land area. And because the S-word accounts for less than a twentieth of a percent of all toponyms nationwide, expressing the resulting rates as *squaw* toponyms per 100,000 provides a more readable map key (fig. 3.7). Although Oregon, with 323 *squaw* names per 100,000, remains in first place, Idaho, with just over 300 per 100,000, rises to second place. By contrast, California—with proportionately fewer pejoratives than Arizona, Nevada, South Dakota, Utah, Wisconsin, and Wyoming—shows up in the map's third category, along with Michigan, Maine, and seven other states. Michigan, with 99 per 100,000, is just ahead of California, while Maine, with 71 *squaw* names per 100,000, has the highest rate east of Michigan. (Maine's rate would be even higher had I considered names based on *squa*, a spelling unique to New England.) It's clear that the West's uneducated miners left a stronger imprint on the macrogeography of *squaw* than the East's and upper Midwest's unschooled lumbermen.

Because lack of a ready replacement obviates blanket renaming and the Board on Geographic Names insists upon historical significance, states willing to rename their *squaw* features must buckle down to the bureaucratic formalities of proposing suitable substitutes. It's not an impossible task, as Minnesota demonstrated in 1995 with a straightforward law proposed a year earlier by two high school students.[53] The statute's assertive and unambiguous wording requires local consultation, places responsibility in the hands of a specific state official, and sets a deadline.

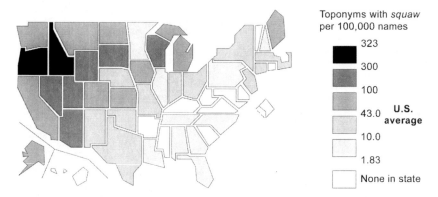

Figure 3.7. Official names containing *squaw* per 100,000 geographic names, by state.

On or before July 31, 1996, the commissioner of natural resources shall change each name of a geographic feature in the state that contains the word "squaw" to another name that does not contain this word. The commissioner shall select the new names in cooperation with the county boards of the counties in which the feature is located and with their approval.[54]

Despite inevitable complaints, renaming in Minnesota has been impressively smooth. Between 1996 and 2000, the federal board approved name changes for eighteen physical features. Substitutions like Nokomis Island (formerly Lone Squaw Island, now based on the Ojibwa word for "grandmother") have a strong native flavor, while others, like Little Woman Lake (formerly Squaw Lake) and Fall Creek (formerly Squaw Creek), have a distinctly non-ethnic aroma. There's even a little subtle humor: although a GNIS history note attributes the substitute for Squaw Lake in Becker County to the Ojibwa word for "flower," Garrison Keillor fans have their own interpretation of the replacement toponym Lake Wahbegon.

Maine enjoyed less success with a 2000 law that tacked *squaw* and *squa* onto a 1977 ban on *nigger*. Although the earlier law prohibited *nigger* as a separate word or part of a word, the amended statute applies only to *squaw* or *squa* used as a separate word.[55] Thus Squaw Point is verboten while Squawhead Point would be permissible. The law requires local officials to find replacements, authorizes the Maine Human Rights Commission to mediate complaints, and provides for court-ordered renaming if all else fails.[56] As I write, Maine has 21 official *squaw* names in GNIS, 14 of them in Piscataquis County, which tried to solve the problem by changing all *squaws* to *moose*. The federal board thought otherwise.[57] Thus far the possibility of judicial involvement seems a hollow threat.

South Dakota's derogatory names law, enacted in 2001, not only targeted 38 specific names but listed specific replacements for 20 of them. To address wider problems as well as win broader support, the legislation revised 11 occurrences of *Negro* as well as 27 *squaw* toponyms.[58] Among other mandated substitutions, Squaw Teat Butte would become Peaked Butte, Squaw Humper Table was to be renamed Two Bulls Table, and Negro Hill was to be called African Hill.

Several months later Oregon banned *squaw* toponyms, or pretended to, with a tortuously worded law in stark contrast to the simplicity of Minnesota's statute.

Be It Enacted by the People of the State of Oregon:

SECTION 1. (1) As used in this section and section 2 of this 2001 Act:
(a) "Public body" has the meaning given that term in ORS 192.410.
(b) "Public property" has the meaning given that term in ORS 131.705.
(2) Except as required by federal law, a public body may not use the term "squaw" in the name of a public property.

SECTION 2. Notwithstanding section 1 (2) of this 2001 Act, a public body that, on the effective date of this 2001 Act, owns or leases public property with the term "squaw" in the name of the public property may use the term "squaw" in the name until the later of January 2, 2005, or two years after the United States Department of Agriculture and the United States Department of the Interior discontinue the use of the term "squaw" in the names of geographic places.

SECTION 3. This 2001 Act being necessary for the immediate preservation of the public peace, health and safety, an emergency is declared to exist, and this 2001 Act takes effect on its passage.[59]

Does the law have teeth? I doubt it. As I read section 1, the statute applies only to public property, and according to section 2, *squaw* names can be used indefinitely if the federal departments of Agriculture and the Interior don't impose a blanket change. A companion bill calls "upon the United States Secretary of the Interior, United States Secretary of Agriculture, United States Board on Geographic Names and Oregon Geographic Names Board to remove the term 'squaw' from names of geographic places in the State of Oregon." Despite five "whereas" clauses, one of which labels the word "derogatory, a racial slur, and as such offensive to Oregonians, Indian and non-Indian alike,"[60] the resolution is little more than a toothless attempt at moralizing and buck-passing. As I write, the four names changed officially, with Washington's approval, since spring 2001 reduced Oregon's catalog of official *squaw* toponyms from 170 to 166.

Idaho's legislature dealt with the *squaw* question by asking the State

Historical Society to hold hearings and develop a plan for changing offensive toponyms. Without mentioning the S-word specifically, a resolution passed in 2002 states a clear goal and acknowledges Washington's role in changing names on federal maps.

Be It Resolved by the Legislature of the State of Idaho:

WHEREAS, Section 19, Article III, of the Constitution of the State of Idaho prohibits the Legislature from passing local or special laws changing the names of persons or places; and

WHEREAS, the state of Idaho encourages tolerance and understanding between all of its citizens; and

WHEREAS, there are certain names of places and geographic areas in Idaho that may be offensive to all or some segments of Idaho's population; and

WHEREAS, it is appropriate for the State Legislature to encourage the eventual renaming of offensive geographic place names in the state of Idaho by encouraging individuals and local units of government to identify and achieve consensus as to geographical place names in the state of Idaho with offensive names.

NOW, THEREFORE, BE IT RESOLVED by the members of the Second Regular Session of the Fifty-sixth Idaho Legislature, the House of Representatives and the Senate concurring therein, that we encourage individuals and local units of government to recommend those geographical place names in the state that are deemed offensive to the State Historical Society, who will adopt procedures to consider these recommendations for changes of names of places and geographic areas that individuals or local units of government deem offensive, taking into consideration such concerns as fiscal impact and whether there is consensus with other individuals or local units of government, and forward the names of places deemed offensive to the United States Board on Geographic Names.[61]

The resolution sets up a process for change and invites persons who feel offended to identify specific toponyms and nominate replacements.

Getting there was not easy. A similar resolution had failed a year earlier after contentious public debate in which native rights advocates stressed *squaw*'s offensiveness to Indians and women of all races, as a crude disparaging synonym for female genitalia, while linguistic conservatives condemned their campaign as political correctness run amok.[62] In convincing the public to look beyond comparatively mild dictionary definitions, the activists first had to win over influential Indian leaders like state senator Ralph "Moon" Wheeler, who became one of the resolution's sponsors. "I grew up thinking this was a name for a female Indian," Wheeler confessed to a reporter for the *Idaho Falls Post Register*.[63] "I guess I'm a little ashamed of that at this point." He made amends by calling *squaw* "one of the most vile, disgusting, crude terms when translated into English . . . a term I would hesitate to use around the campfire with my fishing buddies." Some of his colleagues disagreed vigorously. In early 2001 state representative Twila Hornbeck, who considered the place-names bill unconstitutional, asserted that she "never thought 'squaw' was a bad word [and] still [doesn't] think 'squaw' is a bad word."[64] But a year later she supported a similar measure.

Jeff Ford, chair of the Idaho Geographic Names Advisory Council and a strict dictionarian, was the resolution's harshest critic. No friend of efforts to "sanitize history," he had complained bitterly in September 2001, when the federal board bypassed the council in renaming Chinks Peak (now Chinese Peak).[65] For the council to ponder whether specific toponyms were derogatory to this tribe or that was "bull----," he told a reporter for the *Lewiston (ID) Morning Tribune* a year later.[66] Ford resigned in protest after council members voted ten to eight to accept a proposal from the Nez Percé tribe, which wanted to change Squaw Creek to Waw'aalamnine Creek (meaning "fishing creek") and Squaw Saddle to Wacamyoos Likoolam Saddle (meaning "rainbow ridge"). "I am absolutely convinced that the majority of people in Idaho don't want to change," he opined, "and even if they do change, they won't be able to pronounce them."[67]

Although Ford has a point, the federal board promptly approved the tribe's recommendations. However troubling the new names' pronunciation, they satisfy a board policy requiring an English generic like Creek or Saddle. You might not know how to say a feature's name, but at least you'll know whether it's a stream or a mountain. Because the features are on tribal land, it's unlikely they'll prove a hardship for non-Indians.

Idaho was not the only state in which bills to ban the S-word met strong resistance.[68] Critics often resorted to sarcasm, as in Lake

County, Minnesota, where officials suggested renaming features Politically Correct Creek and Politically Correct Bay.[69] Although some opponents cited the cost of reprinting maps and replacing signs, most reflected a general weariness of political correctness, which strained the limits of newsworthy silliness in the early 1990s.[70] Despite objections, occasionally well-founded, that some *squaw* toponyms never reflected derogatory intent,[71] public sentiment was moved by the obvious distress of Native American women[72] as well as a general malaise about the word's allegedly obscene origin—newspapers wouldn't print it, but an emerging mental image of a translated map showing "Cunt Peak" was repugnant to many non-Indians. Moreover, the argument that renaming features was akin to rewriting history proved particularly vacuous insofar as variants in GNIS and old maps in historical collections provided a enduring record of past cultural landscapes. Politically correct or not, applied toponymy is a political process, like zoning and road construction, and thus at least moderately responsive to public opinion.

Whether Native American activists can make a convincing case for replacing *papoose* toponyms is another matter.[73] Defined as a "Native American infant or very young child," *papoose* seems neither offensive nor latently obscene, and it's unlikely the infants themselves feel any pain from the usage, either as babies or when they grow up.[74] Activists note the similarity to *pickaninny*, which dictionaries identify as an offensive term for young African American children, and point out the absence of an equivalent word for white babies.[75] Even so, their assertion that *papoose* "makes a baby an object" is hardly a persuasive argument for removing it from maps.[76] Although the 137 official toponyms based on *papoose* no doubt contribute in a minute way to the social construction of race and racism, their presence has inspired little action—GNIS lists only eleven variants, with the last replacement officially endorsed in 1969, when Utah's Papoose Lake became Wigwam Lake. While some eradication seems inevitable, particularly on tribal lands, *papoose* place names are markedly less significant, in both number and offensiveness, than names containing *squaw*.

If finding appropriate replacement names is too troublesome for local names authorities, there's an easy universal strategy, which the federal board is understandably reluctant to attempt: when a large group of toponyms proves broadly offensive, simply declare the features "unnamed" until suitable substitutes are found and approved.

Benefits include a more timely recognition of native people's objections and a strong incentive for picking new names. Even so, this ploy would be little more than an administrative Band-Aid insofar as the previous name would no doubt persist in local usage as well as on commercial maps, not to mention on Geological Survey topographic map sheets, most of which are twenty years or more out-of-date—a reality that makes renaming a bit less pressing than it might otherwise appear. In 2003 the board approved a new principle affirming its long-standing practice: "Features once named cannot be unnamed."[77]

CHAPTER FOUR

Body Parts and Risqué Toponyms

Offensive toponyms fall into two categories. One type, examined earlier, denigrates racial and ethnic groups. The other variety, dealt with here, offends folks bothered by rude or otherwise impolite references to body parts, sex, excrement, and other no-no's. A form of geographic cussing, rowdy feature names are markedly less controversial than their ethnically derogatory counterparts, partly because the irreverent miners and ranchers responsible for most of them avoided the F-word and similar shockers, and partly because questionable toponyms occur mostly in remote, sparsely inhabited areas with few eyebrows to raise. Indeed, an outsider who objects to a locally acceptable "naughty name" is quickly branded a stuffed shirt or prude.[1]

Many of the risqué toponyms I have in mind seem more impish than salacious, and some are innocent victims (or beneficiaries) of society's penchant for giving old words new, markedly different meanings. The most famous is Intercourse, Pennsylvania, named in the early nineteenth century to commemorate commerce, not copulation, but long a favorite of souvenir makers.[2] Not far away is Blue Ball, named not for congested testicles but for the Blue Ball Tavern, an inn with a blue ball on its sign.[3] Another Lancaster County inn, with a sign exemplifying the maxim "A bird in the hand is worth two in the bush," inspired the name of Bird-in-Hand, a village gleefully ridiculed as a reference to male masturbation.[4] And then there are ambiguously suggestive names like French Lick (in Indiana), a French settlement near a salt lick patronized by animals craving sodium—toponyms best avoided by list makers wary of being accused of a dirty mind.[5] That's my excuse for citing Lester Dingman, whose list of double entendres includes Coon Butt (in Tennessee), Bloody Dick Creek (in Montana), and Wee Wee Hill (in Indiana).[6] Dingman served a two-year term as executive secretary of the Domestic Names Committee in the early 1970s, and his litany of titillating toponymy probably reflected colleagues' interpretations as well as his own.

Despite these examples, the male anatomy is commemorated far less frequently than the female form because phallic landforms are comparatively rare in nature and most of the namers were men. While I found nothing else evocative of the male member in GNIS— please don't ask what I tried—the database yielded twenty-eight feature names based on *tit*, an even hundred with *nipple*, and a handful based on *teat* or *breast*. Not surprisingly, almost all the features are summits, and few are east of the Mississippi (fig. 4.1), where the early wave of white settlers generally had stronger religious ties.[7] The Maine

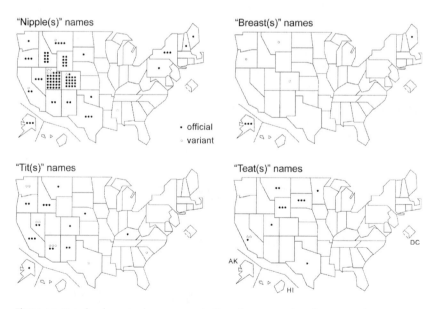

Figure 4.1. State-level counts show a concentration of mammary-based toponyms in the West.

coast boasts two prominent exceptions: a small island with the one-word name Nipple lies less than a quarter mile from a tiny rocky island named Virgins Breasts (fig. 4.2).[8] Given appropriate topographic stimuli, fishermen and mariners can be as toponymically randy as miners and ranchers.

Some namers mixed adolescent angst with a wry sense of humor. Twenty miles east of Walla Walla, Washington, is a pair of closely spaced peaks named Milk Shakes on U.S. Geological Survey topographic maps. Although an online GNIS history note attributes the toponym to "lonely male settlers who felt [the] formations resembled breasts," some of these settlers no doubt preferred the variant name Twin Tits. A similarly suggestive landform eighteen miles northeast of Juneau, Alaska, is unabashedly identified on a 1997 USGS map as Brassiere Hills, an appellation reinforced cartographically by the tree line surrounding the twin summit (fig. 4.3). According to USGS folklore, the name appeared on a 1948 edition of the map but was deleted from a later version because a prudish mapmaker deemed it too risqué.[9] No less a curiosity is Maidenform Peak, a western Wyoming

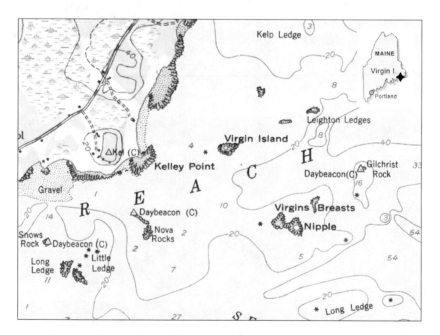

Figure 4.2. The 1977 edition of the Jonesport, Maine, quadrangle map, published by the U.S. Geological Survey at 1:24,000, shows small islands named Nipple and Virgins Breasts less than a mile southeast of Virgin Island.

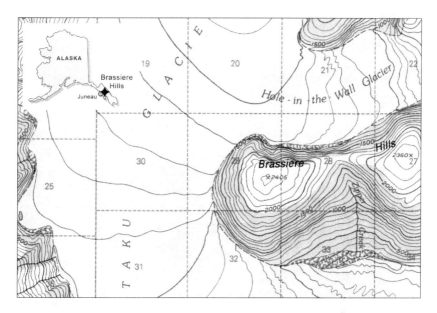

Figure 4.3. Brassiere Hills, as shown on the 1997 U.S. Geological Survey map of the Juneau B-1, Alaska, quadrangle, published at 1:63,360. Image is truncated on the right by the eastern boundary of the map sheet.

feature with a shape-inspired toponym commemorating a particular brand. Although the name connotes a cover-up, there's no prudery here in rural Teton County, home of the evocatively named (if you understand Spanish) 13,770-foot summit Grand Teton.

Names scholars can marvel at the commemorative meaning of *nipple* names, a third of which begin with a personal possessive like Elsies, Marys, Mollys, or Sadies. (The genitive or possessive apostrophe is not normally allowed because the Board on Geographic Names prefers not to show possession.) For whatever reason, Molly outranks all other honorees, with Utah alone accounting for eleven *Mollies Nipple, Mollys Nipple,* or *Molleys Nipple* toponyms. Whether a *nipple* feature tagged with a personal name celebrates the namee as a person or the namee's anatomy is difficult to discern—an anatomical reference seems unlikely for Dans Nipple (in Wyoming) and Peters Nipple (in Utah). Equally intriguing is *nipple*'s geometrically baffling application to lakes (in Colorado and Utah), a spring (in Utah), and a valley (in Colorado). With more time on my hands, I'd enjoy delving into the history of Nipple Church, the variant name of a Mississippi house of worship now officially known as Tabernacle

Church. Although the name might memorialize a steeple, it seems strangely irreverent.

A less humorous aspect of mammary toponymy is the denigration of Native American women by feature names like Squaw Tit, in its singular or plural form (see fig. 1.2). Derogatory intent seems a bit obvious insofar as *squaw* is far more commonly paired with the mildly obscene *tit* than with the more numerous and clinically correct *nipple*. My canvass of GNIS found only two of the latter: Squaw Nipple (in Montana) and Squaw Nipple Peak (a variant for Squaw Dome, in California). By contrast, *squaw* is part of 19 of the country's 28 *tit*-based names (fig. 4.1), and accounts for roughly equal portions of the 19 official names and 9 variants. What's more, unlike the *nipple* appellations that affectionately commemorate white women named Elsie or Molly, none of the *tit* toponyms mentions anyone, white or Indian, by name. And of the six features with *squaw* variants, four still have *squaw* in their official name. Apparently *tit* was more offensive than *squaw* to whoever sanitized the official names of Arizona's Squaw Butte (formerly Squaw Tit), Nevada's Squaw Mountain and Squawtip (both formerly Squaw Tit), and Texas's Squawteat Peak (formerly Squawtit Peak). By contrast, geometry edged out racism when features formerly known as Squaw Tit became Thimble Mountain in California and Pushtay (a Sahaptin Indian word for "small mound") in Washington State. These subtle substitutes suggest a solution for state officials troubled by toponyms pointedly offensive to feminists and Native Americans.

A few feature names allude to what jargon-loving social scientists refer to as the sex trade. Quaintly risqué euphemisms on the cartographic landscape include Cat House Creek (in western Montana, near the Idaho border) and Pleasure House Creek, Pleasure House Point, and Lake Pleasure House (all near Virginia Beach, Virginia). GNIS lists a number of other names containing *pleasure*, but none particularly suggestive of a bordello. Similarly, none of the few dozen names containing *hooker* overtly imply prostitution, not even the two in Nevada.

The most brazen example I found is Whorehouse Meadow, a feature in southeastern Oregon, within the Steens Mountain Recreation Area, run by the Bureau of Land Management (BLM).[10] As names aficionado Lewis McArthur reported in his 1974 book *Oregon Geographic Names:*

In the early days "sin" was considered an unavoidable adjunct of life in the cattle and sheep country. During the summer one or more of the female entrepreneurs from Vale would set up facilities under canvas in this accessible but secluded meadow a mile east of Fish Lake. Houses of this category, wood and canvas, passed with the end of open cattle range and all that remained was the name on the slopes of Steens Mountain. . . .[11]

McArthur's fondness for the bawdy toponym is apparent in his objection to its attempted erasure.

In the 1960s the BLM issued a recreation map and in deference to the moralists substituted a namby-pamby name, Naughty Girl Meadow. The USGS advance sheet of the Fish Lake quadrangle followed suit but in 1971 the [Office of Geographic Names] took strong exception to the change. As this is written, the final decision is pending before the federal arbiters. *O tempora! O mores.*

McArthur's was not the only complaint. In a late 1972 story on the playful pasture's *naughty* name the *Washington (D.C.) Star-News* reported that "folks in the Wild West wish those Puritan pencil-pushers in the federal bureaucracy would leave their colorful place names alone."[12]

There's more to the story than local resistance to priggish mapmakers. According to the Domestic Names Committee case file, a USGS topographer working on the first large-scale map for the Fish Lake quadrangle visited the area in the mid-1960s and inquired locally about the feature, unnamed on existing maps. While the story of mobile madams with tent bordellos catering to summer sheepherders was hardly a secret, all informants called the place Naughty Girl Meadow, which subsequently appeared on the 1968 map (fig. 4.4, *upper*). The mapmaker apparently talked only to park officials, who either didn't know or deliberately suppressed the *whorehouse* version. The Oregon Geographic Names Board became aware of the discrepancy in 1971, and Lew McArthur wrote to the USGS western office in Menlo Park, California. In blaming the name change on the BLM, he opined, "Life in Harney County is rougher than some more urban areas, and some facts of life are accepted without a complete and total moral condemnation."[13]

As reported in the USGS case file, the Oregon board complained to Washington. At its August 1972 meeting, the federal board discussed

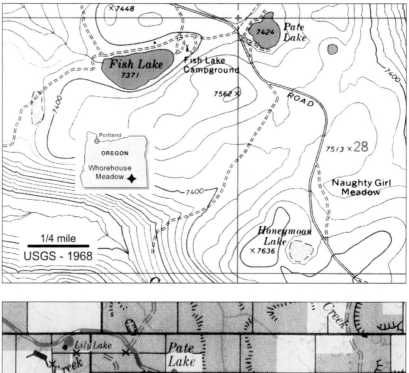

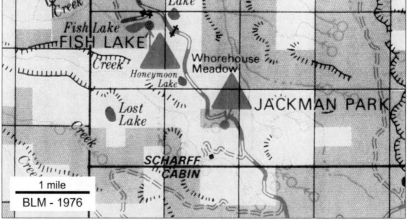

Figure 4.4. In 1983 the Board on Geographic Names replaced the less objectionable name Naughty Girl Meadow, shown here (*above*) in an excerpt from the 1968 edition of the U.S. Geological Survey's Fish Lake, Oregon, quadrangle map, published at 1:24,000, with the original name Whorehouse Meadow, restored on the Bureau of Land Management's Steens Mountain Recreation Lands map (*below*), published in 1976 at 1:158,400 and enlarged to 1:79,200 in the excerpt here.

the issue but took no action. According to the minutes, "The [federal] Board has not received sufficient evidence to warrant a decision, but feels that approval of the name endorsed by the Oregon Board would not be in keeping with its general policy of not approving names which might be considered objectionable by a large segment of the public."[14] In an ironic reversal, the Bureau of Land Management, prudish perpetrator of the *naughty girl* substitute, gave Whorehouse Meadow its cartographic debut on a 1976 recreation map (fig. 4.4, *lower*), even though two years earlier a county map had labeled the feature Naughty Girl Meadow. In 1982 McArthur cited the BLM map in a formal proposal to the Oregon board. "All old local inhabitants agree that Whorehouse Meadow is the old and correct name," he insisted.[15] The state board agreed and sent a formal request to Washington, where nine months later the federal board voted "to give official recognition to a name reported to be in established local use."[16] As in most decisions, the federal board trusted its state counterpart and did not insist on informants' names, affidavits, or other evidence.

Local loyalty to off-color place names is best exemplified by Dildo, Newfoundland, a small fishing village forty miles west of St. John's. Dildo appeared on maps of the area as early as 1771, and the name is not an isolated occurrence: the topographic map covering the village also identifies Dildo Islands, Dildo Arm, and Dildo Cove (fig. 4.5). The name's origin is obscure. Although the *Oxford English Dictionary* mentions a late sixteenth-century term for artificial penis ("dildoe of glasse"),[17] it's unlikely the original namer had in mind the shape or function of the electromechanical vibrator that was invented in 1880s as a medical device, marketed in the early twentieth century as a "personal care appliance," and reintroduced in the 1960s as a sex toy.[18] In 1985 Robert Elford, a villager embarrassed by the connotation, collected nearly four hundred signatures on a petition asking the provincial government to change the name.[19] Elford, who apparently had no particular replacement in mind, backed off after neighbors who liked the name started ridiculing him in public and calling him at home.

Diverse factors account for Elford's failure. His initiative lacked the homophobic imperative behind the renaming of Gayside, Newfoundland (now Baytona), in 1985 or whatever anti-Soviet feeling inspired the renaming of Mount Stalin (a British Columbia landmark now commemorating Don Peck, a highly regarded local conservationist) in 1987. Local residents had few reminders of Dildo's new, potentially

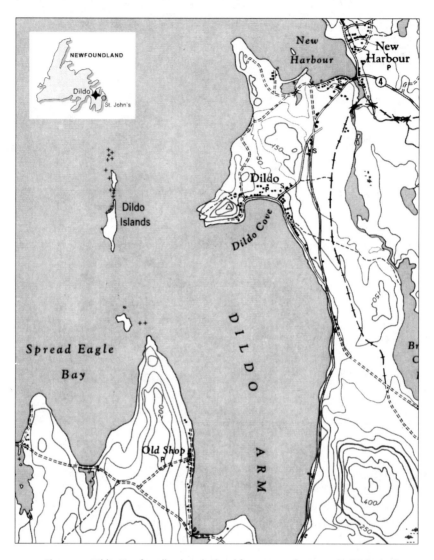

Figure 4.5. Dildo, Newfoundland, and related features, as shown on the Dildo, Avalon Peninsula, Newfoundland, quadrangle map (1N/12), East Half, published in 1952 at 1:50,000 by the Surveys and Mapping Branch of Canada's Department of Energy, Mines and Resources, but reduced here to 1:56,700.

offensive connotation—sex aids were not a regular feature in the news or a lingering icon of cold war rhetoric—and those with a sense of humor could delight in the salacious juxtaposition of Dildo Arm and Spread Eagle Bay (fig. 4.5). Indeed, jokes about the name were a way of being noticed, and perhaps an attraction to tourists who might stop by to mail a postcard or sample local hospitality during Dildo Days, a mid-August weekend featuring "a live band . . . enjoyable games and activities, [and] a beer tent for people 19 and older."[20] The long-standing name was reinforced by its identification of several nearby natural features, and the village had its own postal code (A0B 1P0), which would entail the cost and annoyance of changing one's address. What's more, some Dildodians no doubt felt the same sense of priority as residents of Swastika, Ontario, who resisted the provincial government's renaming their community in 1940 to honor Winston Churchill. Defiantly they ripped down the official sign and put up a replacement proclaiming, "To Hell with Hitler. We had the swastika first."[21]

As with art and literature, questionable but innocent toponyms are easy prey for witch-hunts. A classic example is New York's 1988 effort to ferret out racially derogatory names on state maps. Indignant over the label Negro Marsh (formerly Nigger Marsh) on a state wetlands map, the governor ordered that no state map may "contain any derogatory racial, ethnic or religious name or other epithet" and directed state agencies with cartographic responsibilities to "identify and eliminate problem names from all State maps."[22] Embarrassed that one of its maps had triggered the controversy, the Department of Environmental Conservation took a broad view of "problem names" and compiled a list of forty-five candidates for renaming. In addition to obvious foils like Nigger Pond and Jews Creek as well as names based on ambiguous words like *coon* and *spook,* the list included comparatively innocuous toponyms like Bad Indian Swamp and Christian Hill.[23] Although the rationale for listing the latter pair was dubious but clear, the only reason for including Dingle Hole Wildlife Management Area was a willful misperception that the name referred to *dingleberries,* defined by *Newspeak: A Dictionary of Jargon* as "pieces of excrement clinging to a poorly cleansed anus."[24] Reluctant to get his hands dirty, so to speak, the committee official investigating suspicious toponyms responded with studiously precise bureaucratic prose.

Dingle Hole Road and Dingle Hole WMA do not appear at present to suggest derogatory intent or impact. To the Committee's knowledge the term "Dingle" or "Dingle Hole" does not refer to any ethnic group. If "Dingle" and its variants refer to sexual behavior and/or genitalia, the Committee is not aware of that fact. "Dingle," indeed can be easily traced to the Middle English of the 13th Century and describes a small wooded valley, a dell. A dingleberry is a shrub found primarily in the Southeastern United States. Thus Dingle Hole Road and Dingle Hole WMA may very well function as identifying names with non-derogatory intent. . . .[25]

More blatant scatological toponyms exist, mostly as abbreviations, euphemisms, or unfamiliar spellings. Although one can speculate on the intended meaning of BS Gap (in Arizona) or S.O.B. Hill (in Utah), a GNIS history note leaves no doubt about the variant S H Mountains (now Kofa Mountains, in Arizona):[26]

> Originally named "S H Mountains" by soldiers or miners because of the similarity of lower peaks to outhouses. When ladies came into the area the interpretation became "Short Horn Mountains" in 1900.

Less oblique allusions to primitive plumbing include Outhouse Draw (in Nevada) and Outhouse Creek (in Oregon). If you wouldn't want to drink from or swim in that stream, you'll have no hankering for Shite Creek (Idaho), once you learn (thanks to *Webster's Unabridged Dictionary*) that *shite* is an "obs[olete] var[iant] of SHIT."[27] Equally off-putting are Shitten Creek (in Oregon) and Shitepoke Creek (also in Oregon), insofar as *shitten* can mean "covered with excrement,"[28] while *shitepoke* refers to the heron's "traditional habit of defecating when flying."[29] Although these names might well reflect indigenous words devoid of fecal association, white settlers presumably sanctioned their cartographic inscription, and current residents have no objection.

By contrast, Canadians living on Vancouver Island were more persnickety about carto-cussing, judging from successful efforts to rename Kokshittle Arm (now Kashutl Inlet) and Kowshet Cove (now Cullite Cove).[30] Another Indian name to bite the dust is Paska Township (in Ontario)—although *paska* means "shallow" in Cree, the province changed the name after a local Finnish family objected that a word with a similar sound means "shit" in Finnish.[31] Generalization is difficult, though, because the skimpy, anecdotal evidence is often contradictory. For example, Ontarians asserted a distinct preference for Crotch Lake (renamed Cross Lake in 1941 but restored in the late 1960s), while neighbors of Del Playa Park (near Santa Barbara, California), no doubt stepping lightly, disavowed the variant name Dogshit Park.[32] Is *crotch* acceptably ambiguous but *dogshit* too offensively explicit?

Sex and feces are not the only nonracial themes to incite objections. A toponym that suggests environmental damage can provoke local residents concerned about property values. Thus, while Acid Factory Hollow might be an acceptable name for a valley in remote north-central Pennsylvania, Acid Factory Brook (now Factory Brook) was apparently too off-putting for the citizens of Kent County, Rhode Island, south of Providence. And I'll not be surprised if advocates for people with disabilities start challenging the nation's forty-nine *Cripple Creeks* as demeaning, even though these names usually commemorate injured animals.[33] If names authorities buy their argument, look for vigorous resistance from local residents.

What's clear is that disputes over renaming are as much about control as they are about decency, prudery, aesthetics, or compassion. The same questions arise as when groups or individuals object to racially or ethnically offensive names: who shall name geographic features, and how much clout should local residents have in selecting, defending, or expunging geographic names? Bawdy toponyms survive largely because what's out of place in the Philadelphia suburbs fits in fine in the rugged Rockies, and because the federal board prefers to resist impulsive change and respect local tradition.

CHAPTER FIVE

Going Native

Gaze upward at Devils Tower, in Wyoming, and you'll understand why twenty-three American Indian tribes consider the 1,200-foot pedestal a sacred place.[1] According to native folklore, two young girls were fleeing a grizzly bear when the Great Spirit lifted them to safety by raising the land abruptly, leaving steep walls in which the frustrated bear clawed deep grooves.[2] Geologists tell a different story: the tower is the remnant of an ancient volcano formed 50 million years ago when magma worked its way upward through softer sedimentary rocks and subsequently eroded. As scientists see it, the claw marks are merely vertical cracks between huge columns that formed as the lava cooled. Whatever its origin, the tower is an inviting challenge to rock climbers, thousands of whom turn up each year to the disgust of Plains Indians, who resent the name Devils Tower nearly as much as they detest the desecration of a holy place by mountaineers and tourists. Indians worship at the site every June, and as a former editor of *Indian Country Today* complained, "It's like calling the Vatican the House of the Devil."[3]

Changing the name will not be easy. Plains tribes have different legends and different names for the tower, often more than one. The Kiowa, for instance, call it Tree Rock, while the Lakota Sioux use seven different names, including Ghost Mountain, Grey Horn Butte, Grizzly

Bear's Lodge, and Penis Mountain.[4] The local tourism industry adamantly resists any change, including an informal National Park Service suggestion to keep the name Devils Tower National Monument for the site but rename the feature Bear's Lodge—a disappointing toponym for a spectacular landform.[5] Outfitters and motel operators think it's bad enough that the Park Service now asks visitors to voluntarily refrain from climbing during June. To block a possible name change, they persuaded Wyoming's representatives in Congress to introduce bills banning renaming.[6]

The ploy works because the Board on Geographic Names, according to its bylaws, "will not render a decision on a name or its application if the matter is also being considered by the Congress of the United States or the Executive Branch."[7] No matter that the bill dies in committee—merely proposing a new law can keep the board at bay for two years. Reintroduce a hands-off bill every Congress, and the name Devils Tower is safe.

Wyoming's representatives are following the playbook of Ohio congressman Ralph Regula, who has frustrated efforts to rename Mount McKinley, thousands of miles away in Alaska, for over two decades. Never mind that the Alaskan legislature supports restoring a much older label, Denali, an Athabaskan word for "the high one."[8] Every two years, shortly after a new Congress is sworn in, Regula introduces a bill "to provide for the retention of the name of Mount McKinley." On January 7, 2003, for instance, he filed what became H.R. 164:

> Be it enacted by the Senate and House of Representatives of the United States of America in Congress assembled, That, notwithstanding any other authority of law, the mountain located 63 degrees 04 minutes 12 seconds north, by 151 degrees 00 minutes 18 seconds west shall continue to be named and referred to for all purposes as Mount McKinley.[9]

Similar in wording to earlier stratagems, Regula's proposal was promptly referred to the House Committee on Resources, which had no need to investigate, debate, or vote. Once again the issue was before the Congress, and the federal names board, under a policy formalized in 1981, could do nothing.

Regula is a graduate of the now-defunct William McKinley School of Law and a proud defender of a fellow Republican who lived in his hometown, Canton, Ohio. McKinley represented northeastern Ohio in the House of Representatives 1877–83 and 1885–91 and was elected twenty-fifth president of the United States in 1896. Shot by an assassin in a Buffalo, New York, train station in 1901, he is entombed in the McKinley National Memorial, a Canton landmark. Representative Regula considers him a martyr. To rename the mountain, Regula testified at a federal names board hearing, "would be an insult to the memory of President McKinley and to the people of my district and the nation who are so proud of his heritage."[10]

Unlike most commemorative toponyms, Mount McKinley was branded well before the president's death. William A. Dickey, an Alaskan prospector who admired the Ohioan's spirited defense of the gold standard, named the peak in 1896 to celebrate McKinley's nomination for the presidency—"the first news we received on our way out of that wonderful wilderness," he explained in a January 24, 1897, dispatch to the *New York Sun*, which documented his "discovery" by reproducing his hand-drawn map.[11] According to toponymy guru George Stewart, Dickey's naming "was little more than a joke."[12] As mountaineering historian Terris Moore reported in *Mt. McKinley: The Pioneer Climbs*, Dickey "and his partner fell in with two prospectors who were rabid champions of free silver, and . . . after listening to their arguments for many weary days, he retaliated by naming the mountain after the champion of the gold standard."[13] Although McKinley was not a martyred president when Dickey renamed Denali, the assassination solidified Dickey's frivolous tagging and thwarted later efforts to restore the Athabaskan toponym.

Dickey's naming rights are dubious. The British explorer George Vancouver described the impressive snow-covered peak from a distance in 1794 but didn't name it.[14] An 1839 map by Ferdinand von Wrangel, a Russian naval officer and Arctic explorer who labeled the mountain Tenada, antedates the prospector's drawing by more than a half century.[15] Wrangel's map apparently escaped the attention of U.S. Geological Survey scientist Alfred Brooks, who visited the area in 1902. Acclaimed as the first white man to set foot on the mountain's lower slopes, Brooks reported that "long before Mount McKinley appeared on any map, it already had five names: Denali, Doleyka, Traleyka, Bulshaia Gora (Russian for Great Mountain), and Densmore's Peak."[16] The latter label was favored by prospectors familiar with the vivid descriptions of Frank Densmore, who ventured within sixty-five miles of the mountain in 1889.[17] Brooks's acceptance of the McKinley label reflects the clout of the *New York Sun*, which picked up and promoted Dickey's appellation, as well as the implied endorsement of *National Geographic* magazine, which in 1897 published a short article in which the prospector, a Princeton graduate, invoked the name Mount McKinley eight times.[18] The Geological Survey adopted the name and brought the feature further fame by confirming Dickey's estimate that the peak exceeded twenty thousand feet.[19] By the time journalists and geographers got around to writing about North America's highest mountain, the McKinley toponym was well established.

Alaskans have diverse reasons for resisting Dickey's appellation. Alaskan Indians consider it blasphemous to name any high peak after a person, white Alaskans are miffed about the imperious naming of so prominent a feature after an Ohioan who never visited their state, and historians respect prior discoveries as well as indigenous prerogatives. Although various tribes have used different names for the mountain—GNIS lists twenty-four variants for Mount McKinley, most of indigenous origin—a consensus quickly emerged around Denali as the preferred original name.

An early champion was Hudson Stuck, Episcopal archdeacon of the Yukon. In 1913 Stuck became the first person to reach the summit. In the preface to his book *The Ascent of Denali,* he appealed "for the restoration to the greatest mountain in North America of its immemorial native name." Denali's mislabeling, he argued, was an unfortunate consequence of ignorance and avarice. "No voice was raised in protest, for the Alaskan Indian is inarticulate and such white men as knew the old name were absorbed in the search for gold."[20]

Perseverance triumphed in 1980, when the National Park Service changed the name Mount McKinley National Park to Denali National Park and Preserve.[21] Like other federal agencies, the Park Service can rename its facilities without names board approval, as long as Congress consents. Like many congressional exploits, the name change was inserted without fanfare into otherwise routine legislation, in this case a bill authorizing a significant expansion of the park.[22] In accord with official policy, the Park Service eventually notified the Geological Survey, which conscientiously inserted the new name on subsequent editions of its topographic maps (fig. 5.1). Because the enlarged park is huge—seven times the land area of Rhode Island—its map label typically dwarfs the cartographic tag for its key attraction, Mount McKinley.

State maps could reflect a more radical change. In 1975 the Alaskan

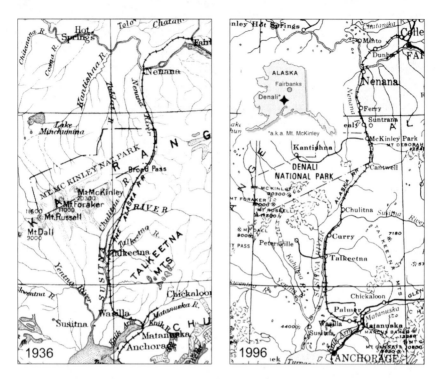

Figure 5.1. Excerpts from 1936 (*left*) and 1996 (*right*) editions of the U.S. Geological Survey's *Map of Alaska*, published at 1:500,000, document the change of Mount McKinley National Park to Denali National Park. But the park's peak attraction remains Mount McKinley.

legislature endorsed Denali as the mountain's preferred name, and a year later the state names board made it official, at least for maps issued by state agencies.[23] Even so, I could not find a single example in our map library's collection of recent Alaskan maps or on the state's Web site. What I found were maps like the online road map (fig. 5.2, *left*), which shows Mount McKinley as the key landmark along the Alaska Range, and the online sectional map (fig. 5.2, *right*) on the "About Alaska" Web page, which touts Mount McKinley as "the highest mountain in North America."[24] No, the state board has not reversed itself. Jo Antonson, who chairs the State Geographic Names Advisory Board, assured me that the official name for the mountain is still Denali. "Technically, maps produced by the State of Alaska should use Denali, but no one enforces this."[25] The state board reviews proposals to send to the federal board but cannot police its decisions. With more

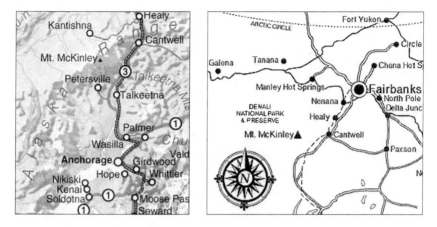

Figure 5.2. Web maps from the Alaska Department of Community and Economic Development reflect a perverse preference for the name Mount McKinley rather than the official name Denali, endorsed for state maps by the Alaskan legislature and the state names authority.

than a century of tradition behind it, the McKinley toponym survives through congressional meddling and cartographic inertia.

Opponents of efforts to restore Native American names for prominent peaks often cite multiple indigenous names as a reason to keep the current toponym. When supporters rally around a single substitute, as with Denali, they'd better prepare for accusations that the proposed name was applied ambiguously to more than one feature. That Denali means "the high one" made it uniquely appropriate for the continent's highest summit—and an unlikely native toponym for lesser summits nearby or even far away. Washington State residents eager to rename Mount Rainier after the city of Tacoma were not so fortunate. Tribes in the region had a variety of distinctive names for the summit, but Tacoma (sometimes spelled Takhoma) was a generic term for any snow-covered peak. In native parlance, Mount Rainier was one of many *tacomas* in the Pacific Coast Range.

Few renaming battles are as complex and sordid as the Tacoma–Mount Rainer controversy, which began in 1883, when promoters eager to make Tacoma the western terminus of the Northern Pacific Railroad relabeled the mountain Tacoma in their ads and brochures.[26] Although Seattle got the railroad, Tacoma chauvinists pressured the state legislature and the federal government for an official name change. Their pleas failed to convince the Board on Geographic Names, which shortly after its formation in 1890 endorsed Captain George Vancouver's 1792 naming of the peak for his friend and fellow naval officer Peter Rainier. However well documented and widely accepted by mapmakers, the toponym struck Tacomans as downright un-American, and they appealed to Congress to change it. According to a bill introduced in 1924 in both the Senate and the House of Representatives, "the man for whom Mount Rainier was named, as the commander of a British ship, engaged in depredations along the Atlantic seaboard in armed opposition to the Government of the United States." In addition to touting Tacoma as the true American Indian name, the proposed joint resolution argued that "perpetuating the name 'Rainier' is contrary to the wishes of the sovereign State of Washington, as expressed in a memorial passed by the legislature of said State in the session of 1917."[27] No mention that the "memorial" was the result of intense lobbying that year by the president of the state senate, the speaker of the house, and the governor, all of whom hailed from Tacoma.

The Board on Geographic Names was not amused. After the House Committee on Public Lands requested a review, board chairman C. Hart Merriam issued a blistering report that condemned the "maliciously false charge that [Rainier] 'ravaged our coasts, burnt our towns.'" Merriam also excoriated pro-Tacoma lobbyists for resurrecting the rumor that the board's original endorsement in 1890 was unfairly influenced by the surreptitious delivery of a carload of Rainier beer, a brand not available until 1894. "Hundreds of people have been misled," he lamented, because of "propaganda emanating from Tacoma and carried on over a series of years, consisting of personal argument and correspondence, newspaper and magazine articles, lectures, and appeals to patriotic and historical societies and women's clubs." Exasperated by pointless arguments, an indignant Merriam charged that "the agitation begun and so persistently carried on by this one city—as against the rest of the world—threatened one of the most firmly established names on the face of the earth and if successful would deal a death blow to the stability of international geographic nomenclature."[28] The joint resolution failed, thanks in part to objections from Seattle residents. But that didn't stifle proposals to rename Mount Rainier. As recent as 1984, Don Orth, executive secretary of the Domestic Names Committee at the time, reported receiving "a few requests every year" to change the toponym to Takhoma or Mount Tacoma.[29]

As the controversies surrounding Devils Tower, Denali, and Mount Rainier imply, indigenous place names ignored or misapplied by white settlers can trigger heated conflicts involving religion, language, orthography (writing systems), multiple native names, tribal sovereignty, historical accuracy, or cartographic tradition. To promote consistency and fairness in resolving the competing claims of multiple stakeholders, the Board on Geographic Names devised a formal policy that acknowledged Native American culture[30] and called for "expert documentation"[31] as well as compliance with established orthographic guidelines, principally the requirement that names be "written in the Roman alphabet as normally used for writing the English language."[32] To avoid confusion about the type of feature named, the policy recommended adding a generic term like *mountain* or *river,* "even though the Native American names may already contain generic elements."[33] Denali seems an obvious exception, but if Tacomans ever had their way, Mount Rainier would probably become *Mount* Tacoma.

A key concern is native sovereignty on tribal lands. Names proposed for features under tribal jurisdiction must meet board guidelines and cannot be approved without consulting the tribal government.[34] Although the tribe does not have the final say on official federal names, its leaders enjoy all the clout accorded a state names authority for features outside tribal boundaries. State officials are not stakeholders in these instances and need not be consulted, but the federal board seeks their opinion anyway because expert advice is always welcome.[35] For a feature that is only partly on tribal land, tribal and state opinions carry equal weight.

For diverse reasons, indigenous groups have influence beyond tribal boundaries. Because the name of a feature not on tribal lands can affect a tribe having a "historic or cultural affiliation" with the area, the Board on Geographic Names will consider the tribe's views either directly or through the geographic names authority of the state in which the feature is located.[36] To determine a possible tribal interest, the board or its state affiliate typically consults "maps provided for that purpose" by the U.S. Bureau of Indian Affairs (BIA).[37] In a similar vein, the board will also weigh the opinions of Native American groups "that do not meet the [federal] definition of 'Indian Tribe'" but can be identified with the help of BIA and state officials.[38] In addition to avoiding new names that might prove offensive or inappropriate, the board seeks a fair and informed assessment of local usage for features with unpublished native and non-native names. Consultation with Native American officials can also facilitate the historically reliable replacement of derogatory names, as in Glacier National Park, Montana, where Dancing Lady Mountain, the English translation of an old Blackfeet name, became the new label for Squaw Mountain.[39]

Consultation is both a courtesy and a stratagem for getting it right. After all, if a proposed name is alleged to be Lakota Sioux in origin, it's wise to ask Lakota officials for their take on usage and meaning. In addition to meeting the board's orthographic guidelines and other requirements, the name should be "linguistically appropriate" to the vicinity—no point in labeling a feature with a strange-sounding name in a language never spoken in the area.[40] Unfortunately, direct submissions involving names on tribal lands are rare, and queries about names on nontribal lands are usually ignored.[41] According to Roger Payne, executive secretary of the board's Domestic Names Committee since 1990, tribal councils typically respond to only 10 percent of the board's queries.[42] Although the board must ensure that a tribe has "considered" any change involving a feature on its own lands, for other features a "good faith effort" to contact the tribe will suffice.[43]

Indigenous languages seem especially prominent on maps of the northeastern states,[44] where white settlers eagerly appropriated euphonious toponyms like Susquehanna and Ontario as well as tongue twisters like Pemadumcook (a lake in Maine) and Chargoggagogg-manchauggagoggchaubunagungamaugg (a lake in Massachusetts).[45] Bay Staters, in particular, seem to thrive on commemorating native peoples displaced by their Pilgrim and Puritan forefathers. In 1997, for instance, the Board on Geographic Names endorsed Peskeomskut Island (meaning "place of great waters" in the language of pre-European Americans who once fished there) as the official toponym for a 492-by-246-foot strip of land in the Connecticut River just north of Turners Falls.[46] Although colonists often adopted indigenous names for natural features, they typically relied on their own language when naming politically important places like towns and forts.[47]

In western states, by contrast, toponyms of Native American origin are generally easier to pronounce than back east because cavalry and settlers imposed English on persons as well as places. Recently endorsed commemorative names that reflect this linguistic imperialism include Bigheart Creek (on the Osage Indian Reservation, in Oklahoma), which celebrates Chief James Bigheart, the first elected chief of the Osage Nation,[48] and Haywitch Creek (in northwestern Washington State), which honors a local American Indian doctor who "reportedly lived to age 130."[49] An exception is Ki-a-Kuts Falls, in Oregon, named after a Kalapuyan Indian chief.[50] Although the state board proposed calling the feature Ki-a-Cut Falls, the federal board adopted Ki-a-Kuts as the official spelling after the Confederated Tribes of Grande Ronde reviewed the proposal at the board's request.[51]

Whether in their original language or as English translations, indigenous names would be more common if Native Americans were less inclined to treat descriptive feature names as privileged information. While some tribes considered toponyms spiritually significant, and thus inappropriate for profane use by outsiders, others simply saw no need to advertise them on maps.[52] Many places went unlabeled because few, if any, nineteenth-century mapmakers were as trusted and thorough as anthropologist Keith Basso, who uncovered history and moral values richly embedded in Western Apache place names passed as stories from generation to generation. As one elderly Apache told Basso, "White men need paper maps; we have maps in our minds."[53] While Native Americans usually had names for minor, visually inconsequential places with religious significance as well as distinctive features like rivers and falls, they often ignored large but vague features like mountain ranges.[54]

Nowhere in the United States are indigenous place names as prominent as in Hawaii, where tourists and newcomers tend to butcher hard-to-pronounce, vowel-rich toponyms like Kapukapuahakea (a cape on the island of Moloka'i) or avoid them altogether. Although the Western imperialists who took over the islands' commerce and government named key streets for themselves and their heroes, native names for physical features survived nicely, no thanks to officials who banished Hawaiian from the schools in 1896. Two years earlier a group of American businessmen and diplomats had deposed Queen Liliuokalani and formed a republic, headed by lawyer Sanford Dole and annexed by the United States in 1898 at Dole's behest. Although more than 90 percent of adult natives could read and write Hawaiian at the time of annexation—an impressive literacy rate for the late nineteenth century—by 1995 barely more than a thousand residents spoke the islands' native language.[55] By contrast, Hawaiian place names proved remarkably healthy. However confusing to mainland visitors, exotic toponyms have become a valued part of the tourist experience.

In the lingo of cultural geographers, this commercial exploitation of Hawaiian feature names and the concurrent suppression of traditional language is a form of "anti-conquest."[56]

For government cartographers touting standardization, Hawaiian is not a map-friendly language. Its comparatively few unique sounds and abundance of homonyms makes context especially important, and its orthography invites tampering by linguistic bigots. Territorial administrators were willing to tolerate native place names but only when rendered in the Roman alphabet, which lacks two crucial diacritical marks, the glottal stop and the macron. Written as a single open-quote mark, as in Hawai'i and O'ahu, the glottal stop (also called the 'okina) is a unique letter in the Hawaiian alphabet. Nearly as crucial is the macron (also called the kahakō), written as a horizontal line atop a vowel to indicate a longer sound, as in Kīlauea Volcano. In addition to disseminating abbreviated spellings and misleading pronunciations, mapmakers sometimes added further confusion by arbitrarily consolidating names with two or more words or by splitting one-word names to separate the generic part, describing the type of feature, from the specific part.[57]

Like their indigenous counterparts on the mainland, native Hawaiians eventually won significant concessions from federal mapmakers. Their victory reflects a resurgence of traditional Hawaiian culture that began in the 1960s and includes native music, dance, and literature as well as spoken language.[58] The University of Hawaii initiated courses in Hawaiian in the 1970s, and precollegiate instruction quickly spread throughout the islands, even at the preschool level. Although a revised state constitution, approved in 1978, recognized English and Hawaiian as the state's official languages,[59] federal mapmakers were constrained by a policy under which "the Board on Geographic Names [did] not approve, for Federal publications, the use of writing marks in the written forms of geographic names derived from the Hawaiian and Native American languages."[60] Change came quickly in the mid-1990s, when the National Park Service asked the Domestic Names Committee to allow the glottal stop and macron on some maps.[61] Under new guidelines approved in 1995, "the presence of diacritical marks, special letters, or symbols will not necessarily bar approval of a geographic name."[62] The policy no longer mentions "Hawaiian" or "Native American" but cautions that names will be reviewed individually. Special spellings must be "consistent with a widely accepted standard orthography" and backed up by "substantial evidence of active local use or acceptance of the name as proposed."[63] And when replacements are authorized, the old versions become variants.

The new diacritical marks policy coincided with a major revision of the island's large-scale topographic maps, updated by the Geological Survey using aerial imagery flown in the mid-1990s. The first batch of official toponyms with Hawaiian spelling appeared on the 1999 *Decision List*, which had 244 newly sanctioned names for Hawaii and only 172 additional names for the thirty other states listed, many with only one or two entries.[64] No Hawaiian feature had been named or renamed since 1989, when the federal board official acknowledged Puu Oo (now Puʻu ʻŌʻo), a volcanic vent in Hawaii Volcanoes National Park formed six years earlier.[65] Between 2000 and 2003, the board approved 2,551 additional Hawaiian names, the vast majority modified only by inserting glottal stops or macrons.[66] For changes involving straightforward insertions, the federal board accepted the wisdom of the Hawaii Board on Geographic Names, which vouched for their correctness. To expedite map revision, the Hawaii board forwarded lists of new names simultaneously to the federal board in Washington and the Geological Survey's mapping center in Denver. GNIS typically reflected the changes in less than a month.[67]

Especially prominent are new spellings for the islands themselves.[68] Hawaiʻi became the official name of Hawaii's largest island in 1999, when the old spelling, Hawaii, joined a list of twenty variants, which includes Ovagi and Owhyee but not Big Island, used widely to avoid confusion with the state name. And Oahu, where Honolulu is situated, became Oʻahu the same year. Similarly, the federal board approved Lenaʻi and Niʻihau in 2001, Kauaʻi in 2002, and Molokaʻi in 2003. Although Hawaiʻian Islands (approved in 2001) is now the official name for the group of islands that comprises Hawaii's five counties, the state's official name will remain Hawaii unless Congress amends the statehood act. A state's legal name is fixed when it's admitted to the union, and neither its legislature nor the federal names panel can change its official toponym.

New editions of Hawaii's large-scale quadrangle maps reflect the relaxed rules. Figure 5.3 illustrates the renaming of a pair of summits commemorating a chief who fell in love with twin sisters, whom he could not tell apart.[69] The area is about twenty-five miles west of Honolulu, on the island of Oʻahu. A 1:24,000 USGS topographic map pub-

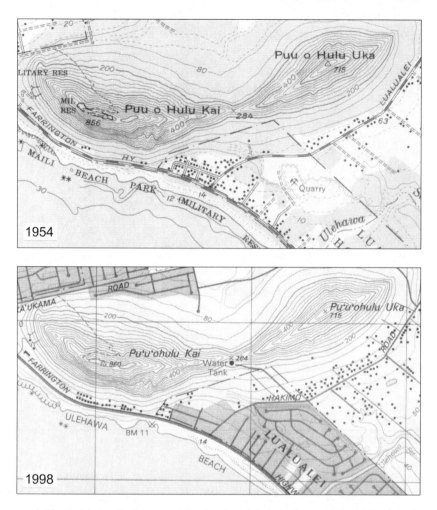

Figure. 5.3. Excerpts from 1954 (*above*) and 1998 (*below*) editions of the U.S. Geological Survey's Waianae, Hawaii (now Wai'anae), 7.5-minute quadrangle map, published at 1:24,000, reflect the recent acceptance of diacritical marks for traditional Hawaiian feature names. The more recent USGS map, revised with aerial imagery flown in 1998, reflects boundary and name changes through 2000, when it was printed.

lished in 1954 identifies the hills as Puu o Hulu Kai and Puu o Hulu Uka (fig. 5.3, *upper*). *Puu o Hulu* refers to the smitten chief, and *Kai* and *Uka* mean "seaward" and "inland," respectively.[70] Identical spellings are found on revised maps published in 1963 and 1983. By contrast, the most recent edition (fig. 5.3, *lower*) uses the labels Pu'u'ohulu Kai and Pu'u'ohulu Uka, which were not approved by the federal panel until

2000, when the map was printed. (Although the map's stated publication date, 1998, refers to the capture date of the aerial imagery used to update roads and other features, boundaries and toponyms were updated through 2000.)

Because of uncertainty, inconsistency, or official changes, it's not unusual to find feature names with three or more spellings, only one of which can be official. An example is the prominent lava flow about two miles northwest of the rim of Kīlauea Caldera, on the island of Hawai'i. A 1997 tourist map published by the National Park Service (fig. 5.4, *lower*) identifies the feature with the two-word label Kīpuka Puaulu, spelled with the macron and endorsed in the 1999 *Decision List*. By contrast, the large-scale USGS quadrangle map published in 1981 employs the two-word version without the macron (fig. 5.4, *upper left*), while the "1995" USGS map—printed in 2003—uses a single-word rendering with the macron (fig. 5.4, *upper right*). GNIS sides with the latter version, Kīpukapuaulu, on the ostensibly older yet actually more up-to-date Geological Survey map. In 2001 the federal names panel amended its 1999 decision after close scrutiny found the split spelling inappropriate and unnecessary.[71]

Pronunciation remains a problem for tourists and new residents, who can get by nicely on older thoroughfares and numbered highways but often become tongue-tied when forced to ask for directions. Visitors committed to getting it right can listen closely to native speakers or consult handbooks like *Place Names of Hawaii*, which includes a pronunciation guide.[72] Or visit http://www.hawaiianlanguage.com, a Web site maintained by volunteers who list common mispronunciations and ask people prone to ignore diacritical marks what they'd think of someone who consistently writes or talks about "Lost Angeles" or "Canda."[73] Clickable sound snippets would be nice, here or on the GNIS Web site, but no one seems eager to undertake the enormous task of finding and recording dependable native speakers. In the 1930s the Board on Geographic Names published a two-page pronunciation key (principally for foreign names), but the current panel steadfastly refuses to enter the thicket of standardized pronunciation, where competing advocates of local dialects would fiercely resist having their way labeled the variant.[74]

New electronic products might someday push standardization along a path today's board is reluctant to explore. An interactive map included with British Columbia's online directory of toponyms dem-

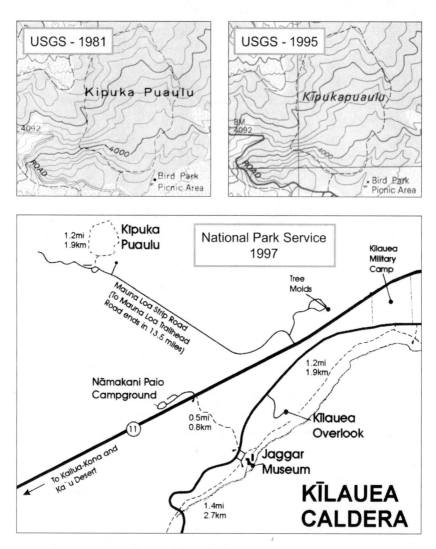

Figure 5.4. Excerpts from the Kīlauea Summit Map (*below*) published in 1997 by the National Park Service, and 1981 (*above, left*) and 1995 (*above, right*) editions of the U.S. Geological Survey's Kilauea Crater, Hawaii (now Kīlauea Crater), quadrangle map show three different spellings for Kīpuka Puaulu, a lava flow two miles northwest of Kīlauea Overlook. The upper-right map, printed in 2003, reflects spelling changes approved after its imagery date, 1995.

onstrates what's technically possible by providing audio clips for thirty-four place names on Nisga'a Lands, a newly empowered aboriginal territory in the northwestern part of the province, near the Alaskan border.[75] Thanks to digital recording and audio plug-ins, native voices can mollify viewers intimidated by spellings like Ts'im K'ol'hl Da oots'ip and its equally arcane phonic equivalent "sim-kohl DAW-AW zip." That this impressively engaging "active map" offers less than three dozen self-pronouncing toponyms attests to the effort required to compile a reliable audio archive.

Another of Canada's First Nations celebrates its cultural heritage with a more conventional cartographic product, the paper atlas. Southwestern British Columbia, near the Washington border, is the geographic focus of *A Stó:lō Coast Salish Historical Atlas,* compiled in eighteen months by a team of social scientists and native activists directed by ethnohistorian Keith Thor Carlson.[76] Unlike regional reference atlases that decompose an area into a quiltlike array of page-size map sheets or cover physical and human geography with a litany of formulaic, region-wide thematic maps, the *Coast Salish* atlas tells the story of the Stó:lō people with forty-six graphic essays, in which carefully juxtaposed maps, texts, tables, and photographs address a specific political, social, or environmental theme. Facsimiles of indigenous or historically significant maps mingle with new artwork addressing contemporary and historical topics at a variety of scales, and the typical map includes an east arrow, pointing toward the rising sun, "the source of all good power." Even so, north is at the top throughout, and three pages of reference maps with official, non-indigenous place names make the atlas accessible to English-speaking readers, including the Stó:lō themselves, who rely almost exclusively on English. As Carlson notes in his introduction, the Haq'eméylem orthography used for toponyms and other terms was devised in the 1970s for the upriver dialect only. Too little time was available for an equivalent treatment of speech patterns found downriver or on Vancouver Island, even though use of the upriver system throughout the atlas might appear to "erase or eclipse more local expressions."[77] As with other facets of applied toponymy, expedient compromise is often unavoidable.

Little compromise was apparent in 1999, when the eastern half of Canada's Northwest Territories became the new territory of Nunavut. A dramatic example of our northern neighbor's commitment to aboriginal self-government, Nunavut is a loose collection of twenty-eight fishing villages with a total population of less than thirty thousand and four official languages: English, French, Innuinaqtun, and Inuktitut.[78] Eighty-five percent of the people are Inuit, formerly known as Eskimos, and their language of choice is Inuktitut, in which Nunavut means "our land." Restoration of native place names became a key element in reestablishing cultural sovereignty, and renaming was well under way in 1987, when the region was still part of the Northwest Territories.[79] Residents of what's now the territorial capital voted 310 to 213 to change their city's name from Frobisher Bay, commemorating explorer Martin Frobisher, to Iqaluit, an Inuktitut word meaning "place of fish."[80] That the vote was not more overwhelming is hardly surprising: many people, even advocates of indigenous home rule, are reluctant to discard familiar place names. What's surprising is the survival, so far at least, of many non-aboriginal names throughout the territory. Of the twenty-eight populated places identified by name on the Canadian government's online map of Nunavut, only eleven appear to have indigenous names.[81] The thoroughness with which Inuit culture reasserts itself should be apparent in the eventual replacement of toponyms like Cambridge Bay and Cape Dorset.

Nowhere in Canada is indigenous renaming as potentially troublesome as in Quebec, which made French the province's sole official language in 1977.[82] Although Canada is officially bilingual, Quebec has been militantly francophone in banning English-only billboards and insisting on a visual presence for French in advertising and product labeling. Although Canada has a federal names board, its provincial equivalents have the ultimate authority over toponyms within their jurisdictions.[83] In Quebec this role falls on the Commission de toponymie du Québec, which compiled a database of over 320,000 names, including street names, in its efforts to standardize spelling and promote the use of correct French, correct English, or correct Inuit. Although the commission has been attacked for trying to suppress English names, largely by translating them into French, its president, Nicole René, notes that even "where English residents have become a minority, or have nearly vanished altogether, English toponyms are being maintained as official when currently used by the present residents."[84] According to René, official recognition of current use has worked well for aboriginal names, which increased from less than sixteen hundred in 1969 to nearly twelve thousand in the late 1990s.

Despite this impressive record, the commission angered Cree traditionalists in 1997, when it celebrated the twentieth anniversary of Bill 101, the most powerful of the Quebec "language laws," by naming 101 uninhabited islands in northern reaches of the province to commemorate noteworthy Quebec writers and poets, almost all francophone.[85] Instead of boorishly applying surnames, names officials canvassed the authors' more notable works for intriguingly evocative toponyms like La Chambre Fermée and Le Desert Mauve.[86] The islands were new features created by a vast hydroelectric project that diverted the Caniapiscau River from Ungava Bay into James Bay. In the eyes of commission bureaucrats, no current names would be affected because their database listed no native names for the area, ironically named Le Jardin au Bout du Monde (the Garden at the End of the Earth). The commission's clever "poème géographique" drew fire from English speakers annoyed by the blatant omission of prominent Anglophone writers who were lifetime Quebec residents as well as from Crees disgusted by the blatant rejection of aboriginal toponymy. It turns out that a consultant to the commission had surveyed Cree settlements on James Bay in the 1970s, but her work did not include inland communities familiar with the flooded area.[87] "All the mountains already have names," maintained Cree elder Samuel Bearskin.[88] "Those names have been there for a long time. There were names wherever I stepped." As far as I can tell, the commission quietly dropped the proposal, thereby sidestepping an equally embarrassing revelation that many of the celebratory "islands" were part of a giant mudflat, exposed annually when Hydro-Québec discharges much of the impounded water through its turbines.

As the American and Canadian examples in this chapter demonstrate, names authorities now recognize the naming rights of aboriginal peoples. Despite noteworthy progress in recent decades, efforts to restore indigenous toponyms are hampered by the rigid rules of standardization, limited budgets, faulty communication with native leaders, and resistance from Euro-Americans demanding equal (or better) treatment for their own traditions and toponyms. Continued conflicts between white and aboriginal place names seems likely as long as cartographic officials insist that no feature may have more than one standard toponym.

CHAPTER SIX

Your Toponym or Mine?

Shortly after I began this book, a two-page letter arrived from Kazu-hiko Koshikawa, director of the Japan Information Center, in New York. Koshikawa didn't say how he found out about my project, but he was eager to contradict "a vigorous public relations campaign" by North and South Korea, which sought to replace "the broadly accepted and internationally well-established name" Sea of Japan with the lo-cationally subjective toponym East Sea.[1] As my research revealed, the campaign began in 1992 as part of a larger effort to eradicate remain-ing traces of Japanese colonial occupation and to stifle recent incur-sions of Japanese popular culture.[2] This resentment is not without cause: imperial Japan wrestled control of Korea from China in 1895, annexed it in 1910, exploited its resources, and forced thousands of its young women to serve as sex slaves in army brothels during World War II. If I were Korean, I'd be resentful too.

A key complaint is that Sea of Japan gained official sanction with-out Korea's consent. In the late 1920s, when the International Hydro-graphic Bureau compiled the first official international list of sea names, there was no Korean voice to contest Japan's recommendation, and Sea of Japan became the worldwide standard.[3] North and South Korea, which rarely agreed on anything, began petitioning for recognition of East Sea in the early 1990s, but Japan adamantly opposed any change.[4] Bureau guidelines apparently permit the listing of multiple names when a dispute cannot be resolved, but agency officials are leery of dual labels because of the confusion that could result.[5] Like passive-aggressive people with weak partners, Japan is winning by refusing to compromise.

Both sides buttress their arguments with historical maps. A twelve-page booklet accompanying Koshikawa's letter cites the Jesuit missionary Matteo Ricci's 1602 world map *Kunyu wanguo quantu* (Complete terrestrial map of all countries), drawn in Beijing in 1602 and lettered in Chinese script, as the earliest map with the label Sea of Japan.[6] Other early examples include Robert Dudley's *Asia carta di ciasete piu moderna,* drawn in Florence in 1646, and Jean-Baptiste Nolin's *L'Asie,* published in Paris in 1704. Nolin saw not one sea, but two, and inserted the label North Sea of Japan between Korea and Japan while describing the waters along the Korean coast as the Sea of Korea, in markedly smaller type. These and other precedents support the consul's argument that the name is unrelated to "Japan's militarist or colonial past."[7] By contrast, a Korean academic who pored over the British Library's map collection for contradictory evidence discovered a 1595 edition of Abraham Ortelius's pioneering atlas *Theatrum orbis terrarum* that lumps the sea between Korea and Japan into a larger China Sea and a 1650 map of Japan that labels the feature Ocean Oriental.[8] More relevant perhaps is the label Eastern or Corea Sea found in *The English Atlas,* published by John Senex in 1711.[9] In the eyes of Korean scholars, these examples undermine Japanese claims of priority.

Koshikawa also contends that replacing Sea of Japan would create confusion because the name is so well established. To bolster this argument, Japanese scholars analyzed 392 maps from sixty countries and found that 97.2 percent of them use Sea of Japan by itself. No map uses East Sea by itself, one uses Sea of Japan followed by East Sea, four use Sea of Japan (East Sea), and six use a different parenthetical supplement.[10] Understandably perhaps, the survey ignored maps from North and South Korea.

Writing off contemporary mapmaking as too politically polluted to help their cause, Korean scholars mined historical map collections for evidence that Sea of Japan is largely an early twentieth-century relic of the Japanese empire. For example, the sea went unnamed on 125 of the 228 pre-nineteenth-century maps of East Asia examined at the

U.S. Library of Congress, and among the remaining 103 charts that identified the water body, only 14 maps labeled it Sea of Japan while another 68 maps called it East Sea, Oriental Sea, or Sea of Korea.[11] East Sea advocates cite similar statistics for map collections at the British Library, the University of Southern California's Korean Heritage Library, and Cambridge University.[12] I'm not certain what, if anything, these numbers prove. While the name East Sea was hardly uncommon, it's easy to rig a map census through selective sampling. I'm especially suspicious of the recent study by Korean historian Lee Sang-tae, who reports that 73 percent of 407 European and American maps published over the last six centuries used either East Sea or Corea Sea.[13]

Aggressive lobbying in the late 1990s won the Koreans some noteworthy victories, proudly touted as a bandwagon well under way.[14] Prominent reference works that inserted (East Sea) below Sea of Japan include the online sources Encyclopaedia Britannica (www.britannica .com) and Microsoft Encarta (encarta.msn.com) as well as the *Oxford Atlas of the World*, the *Rand McNally Premier World Atlas,* and the *National Geographic Atlas of the World.*[15] Unlike many atlas publishers, the National Geographic Society does not issue a new edition every year or two. Although the seventh edition of its world atlas, published in 1999, ignores the Korean toponym, a "patch" (fig. 6.1) posted on its Web site (www.ngs.org) in 2001 accords East Sea a parenthetical mention.[16] As the accompanying explanation implies, the decision followed considerable deliberation.

Figure 6.1. Patch posted on the Internet by the National Geographic Society updates the treatment of Sea of Japan in its 1999 world atlas with a subordinate (East Sea) in smaller type. The original patch, in color, measures 4.5 by 4.0 inches (11.5 by 10.2 cm). From National Geographic Society, "Sea of Japan."

Early in 1999, the National Geographic Society recognized the fact that the South Koreans legitimately disputed the term Sea of Japan. In keeping with the Society's standard place-name convention, we recognize that where a geographical feature is shared by more than one nation, and its name is disputed, we use the most commonly recognized form of the name first and label the disputed name in parentheses. Thus, on our maps, the Sea of Japan appears as the primary label for this feature while the East Sea appears below in parentheses.[17]

Although a few obsessive souls might have cut out the patch and pasted it into their marginally obsolete atlas, inserting a printout between the appropriate pages is a more likely response. In early 2004, the society's Atlas Update Web page carried only twenty-three patches.[18]

Map publishers who found the Korean argument persuasive were no doubt influenced by the persistent letter-writing campaign orchestrated by the Voluntary Agency Network of Korea (VANK). Matt Rosenberg, the geography "guide" (editor) for the About.com Web site, reported receiving repeated e-mails from Korean students who parroted a VANK form letter. Specific complaints focused on About.com's posting of a CIA map of South Korea. When the daily count of e-mail complaints reached twenty, Rosenberg "placed '(East Sea)' on the map not only to stop the attack of extensive bandwidth but also to include the disputed name."[19] Federal agencies are less vulnerable to electronic harassment. At last glance, the Board on Geographic Names, which controls usage by the CIA, is holding the line on Sea of Japan.[20]

While the news media are mostly content with a single well-established name, East Sea occasionally intrudes. In March 2003, when North Korean fighter jets attacked an American reconnaissance plane, the *New York Times* reported the incident as occurring "over the Sea of Japan," which was labeled prominently and without parenthetical supplement in the accompanying news map. Even so, the *Times* quoted a North Korean press release charging repeated prior intrusions "into the air above the territorial waters in the East Sea."[21]

Newspapers face intense lobbying from ordinary letter writers, mostly Korean, as well as embassy officials on both sides. In September 1998, for instance, the *Chicago Sun-Times* published a letter from the Korean consulate general, who argued "the Korean people have never accepted [Sea of Japan]" imposed imperiously "when Korea fell under Japanese occupation."[22] His recommendation that newspapers use both names simultaneously ignored the constraint of limited space on small column-width news maps. And substitution of East Sea as the sole label seems unlikely, at least until the toponym becomes better known. As London's *Independent* noted, "Sea of Japan . . . conveys to the British reader, whose knowledge of geography is in some cases not all it might be, that we are trying to indicate a sea in the vicinity of Japan [while] East Sea conveys no such useful information."[23]

The few newspapers that buy the Korean argument include the conservative *Washington Times,* hardly a bastion of political correctness. The *Times* began using both names "following strong representations from the Korean Embassy in Washington and others." The result was a visit "by a delegation from the Japanese Embassy . . . armed with maps and historical references to argue we should not be using the term 'East Sea' at all." Although editorial policy calls for simultaneous use, occasionally a staff artist omits East Sea. "The last time that happened," foreign editor David Jones noted, "we received a three-page letter from the South Korean Embassy."[24]

If there's an obvious solution, it's buried in a pile of questionable claims. Old maps seem neither relevant nor definitive, while current usage, clearly on the side of the Japanese, ignores the historical reality that toponyms, like boundaries, are political constructions, subject to change. The fact is that no single piece of evidence is overwhelming, and many of the arguments are flawed or pointless. For example, East Sea is not, as the Koreans claim, a neutral term for a body of water east of Korea but west of Japan. Nor is acceptance of the Gulf of Mexico and the Irish Sea by the United States and Britain as germane as the Japanese imply: neither the United States nor the United Kingdom carries a grudge against its less powerful neighbor, and however flattering to Mexico and Ireland, directional references to neighboring states are functionally useful to the Americans and the Brits, and thus deceptively ethnocentric. In this sense, Sea of Korea could serve both nations if the Koreans had been more assertive and the Japanese less intransigent. As long as Koreans insist on reminding themselves and the rest of the world of Japanese atrocities, commercial mapmakers, not compulsively committed to single-name standardization, seem likely to oblige.

A hyphenated toponym like Korea-Japan Sea might seem a logical compromise, but Japanese officials would surely hold out for Japan-Korea Sea. *J* comes before *K*, right? Well, yes, in the Roman alphabet at least, but *J* follows *C* in Corea-Japan Sea, the version Koreans might prefer. The recent controversy over Sea of Japan led some Korean scholars to question the spelling of Korea with a *K*, which Japan is alleged to have substituted in the early twentieth century to gain alphabetical ascendancy over its weaker rival in the 1908 Olympics.[25] This tale seems farfetched, but evidence suggests that Corean Peninsula was at least as prevalent as its *K* counterpart throughout much of the nineteenth century. For example, the map of China in the 1891 edition of *Cram's Unrivaled Atlas of the World* identifies the peninsular province as Corea, an embayment west of the Sea of Japan as Broughton Bay or Gulf of Corea, and a bay north of the Yellow Sea as Corea Bay (fig. 6.2).[26] For the moment, at least, Korean lawmakers are reluctant to incur the cost of reprinting stationery, relabeling embassies and

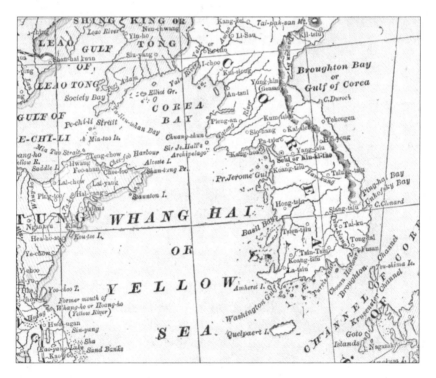

Figure 6.2. An 1891 reference atlas exemplifies the *C* spelling of Korea common in the late nineteenth century. From George F. Cram Company, *Cram's Unrivaled Atlas of the World*, 173.

airplanes, and trying to persuade journalists and mapmakers that the change is appropriate and worthwhile. Despite the expense, a name change is hardly impossible. Immediate acceptance would be unlikely, but official spellings carry considerable clout. In just a decade, China replaced Peking with Beijing.[27]

Approval could come easier in countries with a centralized names authority like the U.S. Board on Geographic Names, which decides how government publications render foreign names. Although evaluating names outside one's borders might seem presumptuous if not unnecessary, federal users such as diplomats, trade officials, and military mapmakers need guidance, especially when disputes arise, multiple renderings occur, or spelling is in doubt. The federal board's first report, issued in 1892, inventoried seventy-eight decisions on foreign names,[28] and a separate analysis of spelling, transliteration, romanization, and similar concerns, published in 1932, listed decisions on twenty-five hundred foreign toponyms.[29] While the 1892 report made no mention of Korea, Japan, intervening waters, or Peking, the much-expanded 1932 roster endorsed Peiping and Korea, with the latter cross-referenced to the Japanese territory named Chosen.[30] Italics identified Corea as a "rejected spelling," while annotations linked Japan to both the Japanese empire and the "sea bet. Japan and Chosen."[31] The list flagged Peking, Pekin, and Peping as improper spellings of Peiping, and denounced Corea and its French equivalent, Corée, as obsolete renderings of Korea.

The federal board now has a Foreign Names Committee (FNC) to complement its Domestic Names Committee (DNC).[32] The National Geospatial-Intelligence Agency (NGA, formerly the National Imagery and Mapping Agency) supports the FNC in much the same way the U.S. Geological Survey supports the DNC. In addition to coordinating deliberations among the CIA, the Department of Defense, and the Department of State (among others), the NGA publishes the *Foreign Names Information Bulletin*, issued irregularly as needed, and maintains the GEOnet Names Server (GNS), an online query system similar to GNIS but less informative and harder to use.[33] With roughly 5.5 million names (including variants) for nearly 4 million geographic features, GNS is nearly twice as large as GNIS.

Before the Internet allowed instantaneous electronic access, the board publicized its decisions mainly through gazetteers that related official names to geographic coordinates and map sheets. Until their declassification in 1968, these lists were largely restricted to diplomats, intelligence officers, and military mapmakers.[34] Sold to the public since the mid-1970s, the federal gazetteers are a useful reference for commercial mapmakers and civilian researchers. Now available in digital form, the series encompassed approximately 5 million names and covered eighty-seven countries as of 1997.

For practical as well as diplomatic reasons, the federal board usually prefers the official local name, except where widespread use of a well-established "conventional" toponym might cause confusion.[35] In these instances, the board often approves both the local toponym and its more familiar equivalent. Thus, Munich and Vienna are allowed, even though the Germans and the Austrians always say München and Wien. Despite official approval of some conventional names, federal agencies set their own policies for choosing an appropriate form. Because what's best depends on the intended audience, a booklet for prospective American tourists would discuss Vienna while a press release to the German-speaking media would always use Wien. In a similar spirit of local accommodation, the board retains diacritical marks like the umlaut in München and allows more than one name for international features like the Danube River (its conventional name), known locally as the Dunaj in Russia, the Duna in Hungary, and the Donau in Austria and Germany. Similarly, features in multilingual areas like Switzerland may have multiple names, like Berne, Bern, and Berna, the respective French, German, and Italian toponyms for the Swiss capital. The board also catalogs official short-form names, like North Korea and Laos, as well as their cumbersome long-form equivalents, like Democratic People's Republic of Korea and Lao People's Democratic Republic.

As indigenous toponyms in the United States and Canada illustrate, names with an unfamiliar pronunciation or spelling typically require some form of conversion.[36] A quick fix for local names with difficult pronunciations, translation often replaces exotically euphonious names like Schwarzwald and Trois-Rivières with more mundane anglicized equivalents like the Black Forest and Three Rivers. Frowned upon as an affront to local custom, translation is less common than transliteration, a more or less letter-by-letter conversion from one alphabetic script to another. (I say "more or less" because an English transliteration might need additional letters if the original alphabet has more than twenty-six letters.) Transliteration works only when sounds are roughly equivalent, as when the Cyrillic Москва becomes Moskva, forerunner of the conventional English name Moscow.[37] The Cyrillic alphabet (thirty-three letters), which looks like Greek (twenty-four letters) to most Americans, could strain the resources and patience of printers and typists, especially in the pre-electronic era. And because Russian, Ukrainian, and other Slavic languages evolved differently, there's more than one Cyrillic alphabet.

When a name is converted to the Roman alphabet, its transliteration is called a romanization. If the original language lacks an alphabet or has a complex writing script like Chinese, romanization might require a sound-by-sound conversion called transcription. Different romanization systems account for the differences between Beijing and Peking, both based, we're told, on the same local Chinese pronunciation. To simplify its work, the Board on Geographic Names, like names authorities in other countries, prefers a single conversion system for each language covered.

American toponymists learned much about romanization from Britain's Permanent Committee on Geographical Names for British Official Use (PCGN), a semi-independent agency established in 1919.[38] After publishing numerous gazetteers for His Majesty's overseas possessions, the PCGN moved on to other areas, particularly in Asia and Africa. In 1945 the agency considered publishing a regularly updated world gazetteer but abandoned the idea because of excessive cost.[39] According to Dick Randall, a former executive secretary of the U.S. Board on Geographic Names and head of its Foreign Names Committee for many years, the American board modeled its postwar gazetteers on the PCGN listings.[40] British influence was apparent as early as 1932, when the U.S. board, "in order to avoid unnecessary duplication and at the same time to utilize the excellent and more voluminous work" of its overseas counterpart, pledged to accept PCGN decisions "in the absence of a specific decision on its own part."[41] A more systematic collaboration arose in the early 1950s, when the federal board and the PCGN began holding combined staff meetings every two years in alternate countries.[42] This collaboration fostered numerous joint romanization systems, which cover twenty-eight languages from Amharic to Uzbek.[43]

A wider collaboration occurs through the United Nations Conference on the Standardization of Geographical Names, held every five years since 1967, and the UN Group of Experts on Geographical

Names (UNGEGN), formed in 1972 to share policies and procedures.[44] Conference delegates representing member nations present papers and vote on formal but nonbinding resolutions advocating domestic standardization and international cooperation. Incisively compelling but seldom implemented as widely as their sponsors might hope, conference resolutions set worthy goals for UNGEGN's nine working groups, in which applied toponymists focus on country names, gazetteers and databases, terminology, publicity and funding, romanization systems, evaluation and implementation, training courses, pronunciation, and exonyms, defined by an official UN glossary as a

> name used in a specific language for a **geographical feature** situated outside the area where that language has official status, and differing in its form from the name used in the official language or languages of the area where the geographical feature is situated. *Examples:* Warsaw is the English exonym for Warszawa; Londres is French for London; Mailand is German for Milano. The officially romanized endonym Moskva for Москва is not an exonym, nor is the Pinyin form Beijing, while Peking is an exonym. The United Nations recommends minimizing the use of exonyms in international usage. See also **name, traditional.**[45]

Formed in 2002 because of growing concern about how mapmakers and authors in other countries render a nation's preferred toponyms, or endonyms, the UNGEGN Working Group on Exonyms is committed to standardizing the spelling of conventional names as well as reducing their use.[46]

As the working group's professed commitment to standardization confirms, exonyms are both useful and inevitable, which might account for their perceived rise in the 1990s after an apparent decline during the 1970s and 1980s, when travelers, writers, and mapmakers became more comfortable with local toponyms or more willing to indulge in the linguistic one-upmanship of calling the Polish capital Warszawa, instead of Warsaw.[47] Confusion arising from inconsistency is an obvious concern, as when one writer says Warszawa, another says Warsaw, and naive readers assume they're talking about two different places. Confusion could prove especially tragic for a small remote town hit by disaster if an international relief agency with an obsolete gazetteer sent supplies to the wrong place. Another issue, a by-product of standardization, is the toponymic sovereignty of a nation that standardizes toponyms domestically and feels entitled to police its exonyms as well. As the UN glossary notes, Beijing, the romanization endorsed by the Chinese government in 1979, is not an exonym. By contrast, the exonym Peking is not only passé but mildly offensive to the Chinese.

Despite UN resolutions calling for fewer exonyms (in 1972 and 1977) or their reduced use (in 1972, 1982, and 1987), names experts recognize the need to distinguish different types of use.[48] According to a 2001 survey, applied toponymists believe exonyms are most commonly used for country names—Holland instead of The Netherlands, for example, or Britain rather than United Kingdom—and are more prevalent in routine conversation, advertising, and the mass media than on maps or for international, scientific, or technical communication.[49] Recommended solutions include appeals to mapmakers and authors of gazetteers, pronunciation guides that make difficult names easier to enunciate correctly, and an international multilingual database that matches outlaw exonyms with their standardized endonyms, officially vetted by the "donor" country for use by "receiver" nations worldwide.[50] Although an easily accessible international database might support efforts to suppress exonyms, officially recognized spellings would appear to sanction their use.

A bigger problem for cartographers is the country that changes its name or renames its cities. While the UN and most national governments willingly acknowledge a country's toponymic sovereignty, commercial mapmakers cringe at a new label that makes their inevitably obsolete atlases and wall maps more so while simultaneously savoring yet another reason why the updated edition planned for two or three years hence can be touted as more up-to-date. In this sense, *glasnost* (openness) and *perestroika* (restructuring), which precipitated the disintegration of the Soviet Union in the early 1990s and the relabeling of its constituent republics, were both a pain and a palliative for atlas publishers.[51] The breakup of Yugoslavia and the reworking of Balkan borders later in the decade had similar consequences, including the tortuous toponym The Former Yugoslav Republic of Macedonia, endorsed by the U.S. Department of State in 1994 (fig. 6.3).

Why not call it just plain Macedonia, like *National Geographic* magazine did in its 1992 map supplement "The New Europe"? Makes sense, I suppose, but not if you want to remain friends with Greeks who consider Macedonia a Hellenic toponym worth defending as

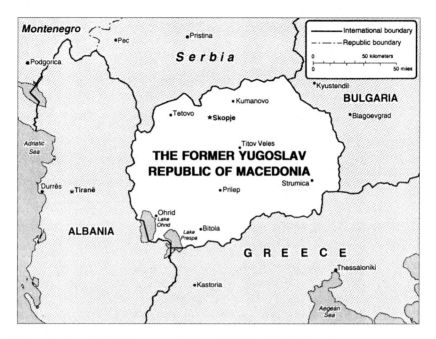

Figure 6.3. From the U.S. Department of State, "Guidance Bulletin No. 12."

vigorously as Disney protects its trademarks. Although Yugoslavian dictator Josip Tito had appropriated the name for a district largely inhabited by Serbs, Macedonia is also the name of Greece's northern province as well as the historic birthplace of Alexander the Great.[52] Among those protesting *National Geographic*'s use of the label was Mark Athanasios C. Karras, Supreme Prelate of the Sovereign and Imperial Order of Saint Constantine the Great. At least that's what his letterhead asserted. Dr. Karras chided the magazine for a "public deception" that "defied national foreign policy as well as international ethics" because the United States and most other nations had not "officially resolved . . . that there actually exists a state of Macedonia."[53] Turns out, *National Geographic* merely reported what our bureaucrats were not yet ready to acknowledge publicly. When the State Department recognized the new country in February 1994, our federal names police tried to placate Greece by approving the awkward abbreviation F.Y.R.O.M. as the only acceptable short form.[54]

Diplomatic posturing is equally apparent when Washington rejects a new name imposed by a regime it considers illegitimate. For instance, in 1989, when the military junta that seized control of Burma the previous year changed the long form from Union of Burma to Union of Myanmar, and the short name from Burma to Myanmar (pronounced Mee-*yan*-mah), American offcials refused to go along.[55] The military regime had allowed elections but refused to cede control to the democratically elected parliament, which rejected the new name. And Washington remains clearly on the side of the largely im- potent Burmese parliament. As the State Department Web site notes, "Due to unyielding support of the democratically elected leaders, the U.S. Government likewise uses 'Burma.'"[56] By contrast, the United Nations is less committed to toponymic politics. Although troubled by human rights abuses there, the UN promptly accepted Myanmar as well as the new name for the country's capital, Yangon, which the United States continues to call Rangoon.[57]

Another cartographic sore spot is the West Bank, named for its east- ern border along the River Jordan and the Dead Sea. The area was part of the Ottoman Empire at the beginning of the twentieth century, part of a British mandate named Palestine after World War I, and part of Jordan in the late 1940s, after the UN created Israel and partitioned Palestine among the new Jewish state and its Arab neighbors. Israel captured the West Bank from Jordan in 1967, during the Six-Day War, and right-wing Israeli cartographers consider it part of Israel. Western mapmakers aren't so sure, largely because of world opinion, a Pales- tinian Arab majority eager for independence, and intermittent peace plans that threaten to expand the jurisdiction of a tragically ineffec- tual quasi-interim government called the Palestinian Authority. Atlas publishers like Rand McNally and George Philip Limited compro- mise with border symbols and labels that confirm the reality of Israeli control and dashed lines or special colors that acknowledge Arab claims as well as most users' understanding that the West Bank, like the Gaza Strip, is different from the rest of Israel.[58] Mapmakers in Is- lamic nations, irate over the West Bank occupation, occasionally retal- iate by erasing the Israeli state, as on a small-scale map in an English- language promotional brochure distributed by the Survey of Jordan (fig. 6.4).

Although political instability is troublesome, mapmakers are equally perturbed when countries like India rename cities with well- established names like Bombay, which became Mumbai in 1995.[59] Users who think they know their geography curse cartographers when they can't find Calcutta (now Kolkata), Saigon (now Ho Chi Minh City),

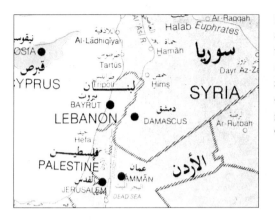

Figure 6.4. Excerpt from a map in a promotional flyer distributed by Jordan's national mapping organization at the 1978 meeting of the International Cartographic Association. Collection of the author.

or Stanleyville (now Kisangani). Blame the confusion on postwar independence, which precipitated the rampant renaming of former British, French, Dutch, Belgian, and Portuguese colonies in Africa, South Asia, and Latin America. Compare a map of Africa for 1950 with a contemporary map, and you'll notice that Nyasaland, South-West Africa, and Upper Volta are now Malawi, Namibia, and Burkina Faso. It's easy to find a dozen more. Less obvious is the slower, more calculated renaming of cities, which created cartographic headaches for years to come, as did countries that changed their names a second or third time to celebrate revolution or a regime change. The former Belgian Congo, for instance, won its independence in 1960, took the name Zaire in 1971, and re-renamed itself the Democratic Republic of the Congo (or simply Congo) in 1997. With no great fondness for dead Belgian monarchs, the Congolese relabeled Leopoldville, their capital, Kinshasa in 1966. Similarly, when white-run Rhodesia (named for explorer Cecil Rhodes) became black-run Zimbabwe in 1980, its capital Salisbury (named for Lord Salisbury, the nineteenth-century British prime minister) was relabeled Harare. And in Russia, where commemorative toponyms are purged whenever an honoree falls from favor, Stalingrad became Volgograd in 1961 and Leningrad was once again St. Petersburg in 1991.[60]

While atlas publishers gave in gracefully if not gleefully, pundit journalists mixed incisive criticism with tongue-in-cheek suggestions for renaming Burma Shave and Peking duck. *New York Times* language guru William Safire proposed "Safire's Law of Nation-Naming: You get only one crack at a new name in each century"—including the label taken up at independence.[61] The *Washington Post*'s Henry

Allen—who noted the popularity of Burma's new name Myanmar among the D.C. dinner party set, where no one calls Norway by its local name Norge—observed that "it is politically correct, and properly multicultural and culturally relativistic, to use whatever name is insisted on by whatever junta or revolutionary council is in power, and the more chaotic, tyrannical and impotent the country in question is, the more correct it seems."[62] And in Britain's *Geographical* magazine, Keith Miller offered an equally astute observation and a dual hypothesis:

> So the general rule seems to be that established countries have no qualms about what they are called; whereas newly-independent states or those that have undergone a major political upheaval feel very strongly about their name abroad. Are they flexing their nationalistic muscles or exhibiting an inferiority complex and a need for reassurance?[63]

However flippant, these quips highlight the toponym's dual role as a convenient verbal reference for geographic location and a powerful symbol of independence and national pride.

CHAPTER SEVEN

Erasures

Disputes more serious than renaming Upper Volta or the Sea of Japan arise when countries covet the same territory, as in Kashmir, torn by overlapping Indian, Pakistani, and Chinese claims. When a sufficiently large map provides a concise description of real and desired boundaries, as on a recent CIA depiction (fig. 7.1), rivals seem not to mind: the map at least registers their claims, and as an apologetic caveat at the bottom observes, "boundary representation is not necessarily authoritative."[1] Hackles rise when a more compact portrait endorses one country's claim or imposes an unacceptable generalized compromise. Microsoft discovered the rhetorical power of maps in the mid-1990s, when India refused to let the software giant import an updated version of its Windows 95 operating system.[2] Microsoft's marketing mavens, it turned out, had overlooked the geopolitical implications of a very small-scale world map displayed during the install to help users identify their time zone. Based on a UN map rejected by the Indian government, the time-zone map distorted the country's outline by omitting the disputed provinces of Jammu and Kashmir—an inflammatory gesture mapmakers at Rand McNally and National Geographic would have known. It was an expensive oversight. In addition to an apology, Microsoft had to replace the offending upgrades with a

Figure 7.1. Excerpt from U.S. Central Intelligence Agency, "The Disputed Area of Kashmir."

cartographically acceptable version. To prevent future losses, the company established a Geopolitical Product Strategy Team to advise on a wide range of cartographic, political, and cultural pitfalls. Who says geography doesn't matter!

Feuding neighbors, especially close neighbors with a history of intense animosity aggravated by differences in language and religion, fight over toponyms as well as borders. And when one group forcibly displaces the other, changing the names of places and geographic features seems a logical strategy for consolidating its grip on new territory. It's an old process, nicely illustrated by Amman, Jordan, which was Rabbath Amman (or Ammon) until the third century B.C., when the Egyptian king Ptolemy II Philadelphus captured and renamed it Philadelphia, a label that held until the seventh century, when an expanding Islamic empire restored the name Amman.[3] Expel the alien, erase his toponymic imprint, and both map and land are yours. But not without consequences: while renaming might blur the claims of displaced peoples and boost the confidence of conquerors, it can heighten the resentment of refugees and remind the rest of the world

of the victor's predations, especially when mapmakers insert parenthetical reminders of old place names. This chapter looks closely at
the ramifications of political renaming in two persistent trouble spots:
Cyprus and Israel.

My first case study looks at the Greek and Turkish Cypriots, who
share the third largest island in the Mediterranean. Most of Cyprus's
900,000 residents speak Greek, write with Greek letters, and observe
Greek orthodox traditions, while a prominent minority numbering
perhaps 200,000 speaks Turkish, uses a twenty-nine-letter roman
alphabet with diacritical marks, and worships in mosques.[4] The most
common second language on the island is English, a holdover from
British administration, which ended in 1960, and a convenience for
British tourists, who contribute significantly to the national economy.
A warm and sunny place admitted to the European Union in 2004,
Cyprus is a popular destination for tourists from other northern European countries.[5] The prospect of increased prosperity through trade
and tourism suggests increased cooperation between ethnic rivals oc-
cupying opposite sides of a variable-width buffer zone known as the
Green Line and patrolled by the United Nations.[6] Running roughly
east to west across an island with a pronounced longitudinal trend
(fig. 7.2), the Green Line separates the Republic of Cyprus from the
self-declared Turkish Republic of Northern Cyprus, recognized only
by Turkey.

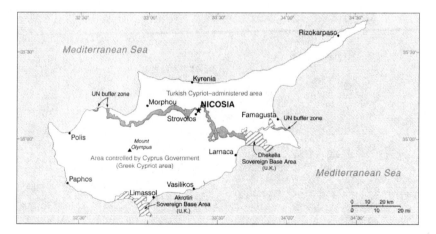

Figure 7.2. CIA map shows the UN buffer zone separating Greek and Turkish Cypriots.
From U.S. Central Intelligence Agency, "Cyprus."

Concentration of Turkish Cypriots in the north reflects the island's location forty miles (64 km) south of the Turkish coastline. Britain acquired the island from the faltering Ottoman Empire in 1878, ostensibly to protect the Ottoman Turks from Russia, and in exchange for turning over the government to a bi-ethnic democracy in 1960, retained two large military bases, on the southwest and the southeast coasts.[7] Many Greek residents resented a constitutional provision that gave Turkish Cypriots, who made up a fifth of the island's population, a disproportionate share of legislative seats and ministerial positions. Although Greek and Turkish Cypriots were unified in their desire for independence from Britain, lack of a distinctive Cypriot nationalism made the new republic vulnerable to separatist agitators, whose violent clashes led to the establishment of a UN peacekeeping force in 1964. Greek Cypriots desired union with Greece—the island was ethnically Greek for three thousand years, even under Ottoman rule— while Turkish Cypriots favored partition into Greek and Turkish provinces.

The Turkish minority got its wish, sort of, in mid-July 1974, when the military junta that had taken over in Athens encouraged a coup d'état by Greek Cypriots in the National Guard, and Turkey sent 40,000 troops to protect Turkish Cypriots.[8] Although the presence of UN peacekeepers and British troops at the two military bases averted a full-scale war between Greece and Turkey, the mid-August ceasefire left Turkish forces in control of 37 percent of the island. De facto partition followed when 180,000 Greek Cypriots quickly relocated south of the Green Line, and 60,000 Turkish Cypriots moved north.[9] By 1990 only six hundred Greeks were living in the north, and a mere hundred Turks remained in the south.[10]

While both sides consolidated their holdings, Turkish Cypriots commenced a campaign described by London's *Financial Times* as "an uncompromising policy of Turkification,"[11] censured by the UN as "a form of colonialism and [an] attempt to change illegally the demographic structure of Cyprus,"[12] and denounced by politicians in the south as "cultural genocide."[13] Although prone to overstatement, the Republic of Cyprus's Press and Information Office railed against a three-pronged strategy that included stationing 35,000 Turkish troops on the island, importing 115,000 Turkish settlers, and trying to "eradicate every trace of a 9,000 year old cultural and historical heritage."[14] Renaming leads the list of cultural assaults.

All Greek place-names have been replaced by Turkish ones. Churches, monuments, cemeteries and archaeological sites have been destroyed, desecrated or looted. Priceless religious and archaeological treasures, part of the world's cultural heritage, are being stolen and smuggled abroad, and illegal excavations and dealings in antiquities are taking place.[15]

Journalists confirmed the charges, at least anecdotally. A September 1976 article in the *Economist* reported, "Towns and villages such as Kyrenia and Lapithos which stood empty and ghostly until April are now filled with settlers, some Turkish-Cypriot, some mainland Turks."[16] A 1997 story in the *Jerusalem Post* described the extent of Turkish "ethnic 'purification' of the north":

Masses of antiquities, including priceless whole collections, have been looted, broken up and sold from the north, church icons included. Irreplaceable painted churches from early Christianity have been hacked to rubble or ruin. The Cypriot government has raced around the world trying to intercept and buy back the most precious items and halt sales of stolen collections.[17]

In a 2003 article in the *Guardian*, reporter Paul Hamilos described a visit to Gypsos, his father's birthplace, now called Akova.

. . . The cemetery was desecrated during the invasion, though some attempt to restore the gravestones has been made. . . . The Church of St. George, where my parents were married in 1973, has been converted into a mosque, its bells replaced with speakers for the call to prayer.[18]

Greek Cypriots were not the only victims. A 1999 London *Independent* report on the "indignities" suffered by Maronite Christians, a tiny Catholic minority affiliated with Rome, described the atrophy of Kormakiti, a village founded seven centuries ago but renamed Korucam as "part of . . . Turkification."[19]

Maps attest to the toponymic purge.[20] Compare the Cyprus excerpt from a 1994 Turkish government tourist map (fig. 7.3) with the more recent map in the CIA's *World Factbook* (fig. 7.2). That the American government does not endorse Turkish renaming is apparent in the disparities for Güzelyurt, Girne, Gazimagusa, and Dipkarpaz, the Turkified replacement names for Morphou, Kyrenia, Famagusta, and Rizokarpaso. Oddly, both the Greek and the Turkish Cypriots adopted new

Figure 7.3. Cyprus excerpt from Republic of Turkey, Ministry of Tourism, "Turkey Tourist Map."

names for the capital, Nicosia, which straddles the Green Line. Equally puzzling, the new names are surprisingly similar, the Greek Cypriots having made Nicosia an exonym by restoring the city's tenth-century Byzantine name, Lefkosia, which their Turkish neighbors render Lefkosa. Intriguingly, the Turkish tourist map includes parenthetical Turkish toponyms for the historically Greek names of three cities along the island's southern coast, but except for Paphos (Baf to the Turks), the Greek and Turkish versions have similar spellings. The map further reveals its persuasive role by highlighting tourist connections with the Turkish mainland and elevating the northern zone to the status of a "Republic" while merely acknowledging the "Greek Cypriot Administration" in the south.

Geographers Sarah Ladbury and Russell King, who mapped name changes across northern Cyprus, uncovered a concerted effort to make Turkish Cypriots from the south feel at home by removing visible signs of Greekness.[21] For every Turkish Cypriot who moved north, three Greek Cypriots fled south, leaving whole villages vacant or largely so, and thus ripe for resettling refugees from the south as well as Turks from the mainland. Before 1974 almost all villages in the north had a Greek name, partly because British mapmakers preferred

Greek toponyms (with an English spelling, of course, for easier pronunciation). In addition to giving Greek villages new names and abandoning the Greek toponym of any village with names in both languages, the government in the north restored the traditional Turkish spelling for Turkish villages like Geunyeli, now spelled Gönyeli. In other cases, a Greek name was merely translated into Turkish. For example, Morphou and its Turkish replacement Güzelyurt both mean "beautiful home."

Transplanted communities, relocated as a unit to preserve kinship ties and social networks, occasionally requested a name commemorating their old village in the south. After refugees from Erenköy were resettled in Maltepe, the new name conferred on the formerly Greek village of Yialousa, the Turkish Cypriot Resettlement Authority agreed to rename it Yeni Erenköy, meaning "New Erenköy." Not all requests were honored: although the new occupants of Akova (known as Gypsos when Paul Hamilos's parents were married there in 1973) all came from a village with the Turkish name Vuda, the government refused to change Akova to Yeni Vuda. To advertise their resentment, defiant residents painted their preferred toponym on walls throughout the village.

Ladbury and King drew an important distinction between the policy of erasing Greek place names, which was "intensely political," and the new names, which were "not political statements in themselves."[22] They found only one exception, in the village of Karaoglanoglu (formerly Ayios Gheorghios), relabeled to commemorate a Turkish officer killed during the 1974 invasion. By contrast, the government in the south never acknowledged the new Turkish names on its maps or road signs, which still point to Kyrenia and Famagusta. This aversion to renaming includes the many new settlements built to house the south's surplus refugees. Wary that an alphabetic name, even the name of a village abandoned in the north, might imply a permanence that Greek Cypriot leaders were unwilling to concede, officials identified refugee settlements only by area and number, as in "Nicosia 3." In much the same way that new Turkish names in the north meant the Greeks weren't coming back, sterile temporary names in the south gave refugees hope of someday returning home.

Political renaming was nothing new for the Turks. In 1924 the Turkish government reacted to fear of growing numbers of Kurds in its eastern reaches by banning the use of Kurdish in schools and newspapers and replacing Kurdish toponyms on maps and road signs.[23] While this earlier purge of place names was perhaps no worse than post–World War I erasures elsewhere in Europe and Asia, the recent episode seems unusually harsh for the late twentieth century. At least that's the opinion of Israeli toponymy expert Naftali Kadmon, a professor of geography at the Hebrew University of Jerusalem and a long-standing member of the UN Group of Experts on Geographical Names. His lively book *Toponymy: The Lore, Laws and Language of Geographical Names* includes a global assessment of contemporary political renaming and credits Turkish Cypriots with "the most extreme form of verbal toponymic warfare."[24]

Perhaps out of patriotism, Kadmon says little about Israel's use of toponymy as a weapon. While his anecdotes from UNGEGN meetings and the quinquennial UN Conference on the Standardization of Geographical Names liken the dispute between Greek and Turkish Cypriots to the conflict between Israelis and Arabs, his critical spotlight falls on the Jordanian cartographers who erased Old Testament names from the map of the Holy Land and their Palestine Liberation Organization counterparts who produced a Palestinian map series on the cheap by photocopying Israel's 1:100,000 topographic maps, replacing Hebrew names with Arabic names, and substituting the PLO logo for the Survey of Israel copyright notice. Despite his conspicuous silence on Israeli renaming, Kadmon candidly concedes as "a 'fact of life' that geographical names and their manipulation can be used as arguments by either side in a political conflict."[25]

His countryman Meron Benvenisti is less circumspect. An author widely respected for his insights on Arab-Israeli tensions, Benvenisti begins his book *Sacred Landscape: The Buried History of the Holy Land Since 1948* with a detailed account of a toponymic struggle with roots reaching back more than two millennia. Where Kadmon dismisses complaints that Israeli maps suppress Arab names with the valid observation that "all Arab towns and villages in Israel carry their Arabic names, and these appear in official Israeli maps,"[26] Benvenisti draws a less kindly conclusion: while renaming elsewhere in the world largely restored names that "had existed for countless generations [with] each ethnolinguistic group having its own version . . . only in Israel was a new toponymy imposed by an official naming committee, which invented most of it."[27]

While the current Arab-Israeli conflict can be traced to the late nine-

teenth century, when Zionists urged European Jews to escape perse-
cution by returning en masse to their biblical homeland, the toponymic dispute began in earnest in the 1920s, when Palestine was a
British mandate, established at the Paris Peace Conference to admin-
ister the former Ottoman territory and foster Jewish immigration
under the Balfour Declaration of 1917. The British administrators
made detailed, systematic mapping a priority and set about compiling
an official list of Arab, English, and Hebrew names for readily identifiable places, including Zionist settlements. (Most places had an Arab
name, many had a Hebrew name, some had both, and all were now to
have an English name.) To ensure accuracy, officials asked the Jewish
Society for the Study of the Land of Israel, a volunteer Zionist group
whose membership included experts in local history, to suggest, correct, or approve Hebrew toponyms for 360 places and historic sites
on a draft list.[28] Because many of the locations were Arab villages, the
society supplied, corrected, or approved only 145 Hebrew toponyms,
all of which were included in the *First List of Names in Palestine*, published in 1925 with the approval of Britain's Permanent Committee on
Geographical Names.

A second, much longer list appeared in 1931. Compiled by the
mandate's education department without consulting either the Jewish
Society or the PCGN, the tediously titled *Transliteration from Arabic
and Hebrew into English, from Arabic into Hebrew, and from Hebrew into
Arabic, with Transliterated Lists of Personal and Geographical Names for
Use in Palestine* angered the Jewish community by replacing Hebrew
names from the 1925 list with Arab names transliterated into Hebrew.[29] Thus, instead of Ashqelon, an established Hebrew name with
a biblical pedigree, the new list called for Isdud, a transliteration of the
Arabic toponym 'Isdud. The General Council of the Jewish Community of Palestine denounced this "bastardization" as "a crass offense to
the Hebrew language," and the PCGN joined the protest with its own
complaints, as did several government departments and even members of Parliament.[30] Aware that Hebrew toponyms were linguistically
and historically complex as well as politically sensitive, the British administration withdrew the 1931 list, reaffirmed its endorsement of Hebrew names in the 1925 roster, sanctioned any Hebrew name used by
generations of Hebrew speakers, and made the Place-Names Committee of the Jewish National Fund the unofficial authority on Hebrew
names in Palestine and a strong influence on the systematic naming

of new settlements.[31] When Israel achieved independence in 1948, many members of the National Fund committee were appointed to the official Israel Place-Names Committee established by the prime minister's office.

Not surprisingly, nation-building is a dominant theme in Israeli applied toponymy, with Zionist symbolism particularly prominent before 1948, and names connoting connections with the biblical past comparatively common after statehood. Geographers Saul Cohen and Nurit Kliot, who studied the ideology of Israeli place names, classified 889 Hebrew toponyms, mostly for villages, towns, and cities.[32] Among 346 places established between 1880 and 1948, 29 percent of the names had an ancient biblical or Talmudic association, 23 percent honored Zionist leaders or philanthropists, and 8 percent commemorated military victories, war heroes, and the like. For the 543 places established between 1949 and 1979, the proportions of ancient and military-heroism names rose to 46 and 10 percent, respectively, while the share for the national Zionist category dropped to 12 percent. According to Cohen and Kliot, this shift reflects a perceived need after 1948 to emphasize continuity with the Old Testament Kingdom of Israel and a diminished desire to commemorate Zionist benefactors.

A new pattern of naming followed the Six-Day War of 1967, when Israeli forces took control of the West Bank, the Gaza Strip, and the Golan Heights, as well as the Sinai, returned to Egypt in 1982. As part of "Greater Israel," the new "Administrative Territories" fulfilled the right-wing Likud Party's ambitions for a more defensible perimeter and afforded an opportunity for establishing new settlements, all requiring Hebrew names. Ancient biblical themes remained popular, accounting for 36 percent of the 174 place names established in occupied areas between 1967 and 1988, while national Zionist and military-heroism themes dropped to 7 and 6 percent, respectively, perhaps to avoid antagonizing the new settlements' Arab neighbors—as Cohen and Kliot point out, most of the new settlements are in areas densely populated by Palestinians.[33] By contrast, abstract names like Orim (meaning "lights") or 'Alumim ("youth")—which connote joy, stability, or confidence—account for 22 percent of the new toponyms, up from 15 percent for "Old" pre-1967 Israel. Instead of symbolizing connections with Israel's historic past, abstract names signify change and optimism.

Maps reveal a striking contrast between the occupied territories, where most Arab names hang on nicely, and pre-1967 Israel, which witnessed a wholesale eradication of Arab toponyms. Meron Benvenisti, who considers nationalist renaming arrogant and dishonest, condemns the Israel Place-Names Committee for endorsing Hebrew names retroactively in the early 1950s while rejecting Arab toponyms made official decades earlier by the British. In chiding Kadmon for "resort[ing] to half truths"[34] by asserting that "a governmental naming authority had never been set up in this country,"[35] Benvenisti notes tartly, and with measured sarcasm, that "the British Mandatory governmental authority, which has scrupulously attended to the standardization of names—perhaps even with excessive zeal—never 'existed,' in the opinion of this Israeli scholar, since it was a colonial authority rather than a Jewish-Israeli one."[36]

In their rush to resurrect biblical place names, Israeli toponymists occasionally labeled uncertain locations. Finding Old Testament places on the contemporary landscape is difficult and potentially controversial, especially when the name disappeared from local usage and the biblical reference is vague. Benvenisti's examples include Yotvata and Mount Hor, ancient names from chapter 33 of the book of Numbers. A tourist pausing at Yotvata "on the road to Eilat, is convinced that he or she is actually stopping at the site of one of the encampments of the Children of Israel during their wanderings in the desert."[37] Disclaimers hidden in obscure documents don't make these careless appellations any less misleading. A particularly controversial example is the placement of Mount Hor, where Moses' brother Aaron was buried, in the central Negev, an arid area southwest of the Dead Sea. Although traditional Jewish and Muslim texts place Aaron's burial site in ancient Edom, within present-day Jordan, Zionist mapmakers wanted it on Israeli soil. When scholarly evidence made this appellation untenable, mapmakers relabeled it Mount Zin but left the name Mount Hor on the map in parentheses.[38]

A different kind of controversy arose in 1939, when Tel Aviv's leading rabbis complained to the National Fund committee about the cartographic commemoration of ancient Sodom by the toponyms Sodom Colony and Sodom Workers Camp. Angered because Sodom connoted homosexuality, the chief rabbis insisted that "this ugly name . . . be erased immediately from our maps and our children's lips."[39] Whether their complaint prompted a relabeling is difficult to determine: neither name appears among the 7,700 toponyms in the 1983

Gazetteer of Israel compiled by U.S. Defense Mapping Agency from official Israeli sources, but there's a Sodom listed as a variant of the official spelling Sedom.[40] However compelling this cartographic inscription on the southwest shore of the Dead Sea, few archaeologists accept it as the historic site of the sinful city destroyed in the book of Genesis.[41] This is just another example of the Zionist obsession with making the Hebrew map look biblical.

Israeli renaming also reflects wholesale abandonment of Arab villages in 1948, when the British pulled out, Israel's neighbors invaded, and the better-trained Israeli army captured considerable Arab land.[42] In dissolving the British Mandate, the UN had partitioned Palestine into interlocking Arab and Jewish states, separated by a long, awkwardly porous border. Determined to defend its new, shorter perimeter, the Israeli government expelled Arabs who had not fled and demolished entire villages, expunging their Arab names from the cartographic landscape. In many cases, leveled areas became fields or orchards for Israeli farmers; in others, an abandoned site proved an advantageous location for a new settlement, with a Hebrew name naturally. Israeli historian Benny Morris documented the process in his book *The Birth of the Palestinian Refugee Problem, 1947–1949*. A pair of dot maps inventory 369 Arab villages abandoned in 1948–49 because of fear, military attack, coercion by Jewish forces or Arabs, or "psychological warfare," and 186 Jewish settlements established during the same period.[43] Although the second map has fewer dots, its pattern is remarkably similar. In early 1949, as Morris observed, "the political desire to have as few Arabs as possible in the Jewish State and the need for empty villages to house new Jewish immigrants meshed with the strategic desire to achieve 'Arab-less' frontiers."[44]

To comprehend what happened, I visited my university's map library, located the drawer with topographic maps for what was Palestine and is now Israel, and set about comparing pre- and post-independence landscapes. Examples of abandoned villages were easy to find. The one I chose includes a small mesa-like feature marked by hachures (lower left of figs. 7.4 and 7.5) and labeled Tall Kisan by the British and Tel Kison by the Israelis.[45] (Tel, the map key notes, means "hill" or "ruin.") This landmark and a pair of intersecting highways made it easy to extract corresponding excerpts, which I enlarged to focus on details easily lost among grid lines, elevation contours, and topographic symbols.

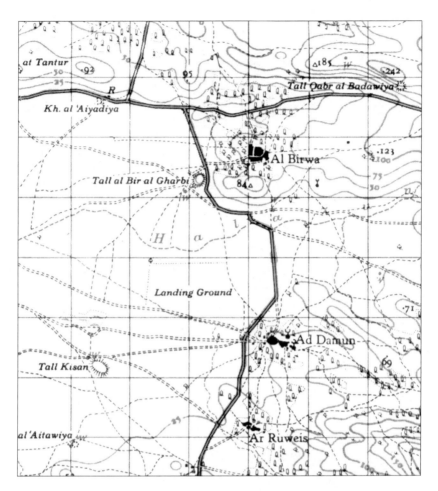

Figure 7.4. The vicinity of Tall Kisan and the village of Al Birwa as portrayed on Sheet 2, Haifa, of the 1:100,000 Map of Palestine, published by the Survey of Palestine in 1943. Excerpt has been enlarged to approximately 1:62,500. Grid lines 1 km (0.76 mile) apart afford a sense of scale.

The area is about ten miles northeast of Haifa. Three villages shown as mere dots on Morris's first map appear as dark splotches on the 1943 Survey of Palestine map (fig. 7.4). Narrow road symbols partition the largest village, Al Birwa, into three irregularly shaped blocks of densely packed masonry buildings. Bushy, treelike symbols immediately west represent orchards, while narrower vertical tree symbols on the other sides portray olive trees. Two miles south, Ad Damun has a smaller cartographic footprint with olive groves on the north,

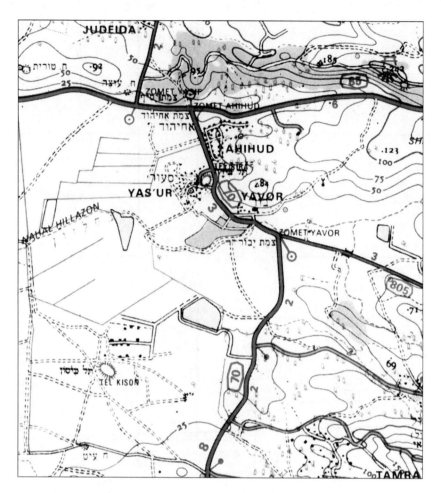

Figure 7.5. Tel Kison and the site of the abandoned village of Al Birwa as portrayed on Sheet 3, Hefa, of the 1:100,000 Map of Israel, published by the Survey of Israel in 1988. Excerpt has been enlarged to approximately 1:62,500 and cropped to correspond to the area shown on figure 7.4.

west, and south. As with Al Birwa, dashed lines show footpaths leading eastward toward higher ground. The double-dashed lines heading north and west are unpaved roads leading downward to the Mediterranean Sea's comparatively flat coastal plain, a suitable site (the map reveals) for landing small aircraft. A mile farther south is Ar Ruweis, with a slightly smaller cartographic presence in a similar topographic setting. All three villages are on the lower eastward flanks of rugged terrain rising to the west. Abrupt turns along the paved north–south road reflect a compromise between minimizing gradient and serving foothill villages.

That was then. Although the road network and some less conspicuous topographic clues helped me locate their sites on the 1988 Survey of Israel map (fig. 7.5), cartographically the three villages simply disappeared. New paved roads run east to west across the sites of the two smaller villages, while Al Birwa is now just a vacant spot on the map, shunned by the planners who laid out Ahihud, a new settlement of detached buildings just downhill and closer to the main road, now Route 70. Missing symbols indicate that dirt roads and footpaths also dropped off the map after the Arab villagers left. According to a table accompanying Morris's map, villagers abandoned Al Birwa around June 11, 1948—a question mark concedes uncertainty about the date. The code for "decisive cause" attributes their exodus to a "military assault on the settlement by Jewish troops."[46] Similar threats emptied Al Birwa's smaller neighbors on July 15–16.

Gone but not forgotten, all three villages survive in cyberspace at PalestineRemembered.com, "the home of all ethnically cleansed Palestinians." Alphabetical and cartographic indexes point to descriptions of abandoned villages and their former inhabitants. The index for the District of Acre (fig. 7.6) shows al-Birwa, al-Damun, and al-Ruways as dots 23, 24, and 26, identified below in romanized Arabic.

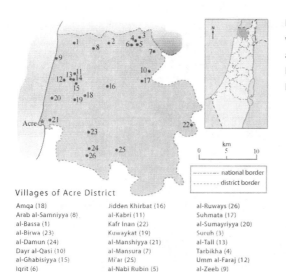

Figure 7.6. The three Arab villages in figure 7.4 carry on as dots 23, 24, and 26 on the District of Acre map from PalestineRemembered.com.

Villages of Acre District

Amqa (18)	Jidden Khirbat (16)	al-Ruways (26)
Arab al-Samniyya (8)	al-Kabri (11)	Suhmata (17)
al-Bassa (1)	Kafr Inan (22)	al-Sumayriyya (20)
al-Birwa (23)	Kuwaykat (19)	Suruh (3)
al-Damun (24)	al-Manshiyya (21)	al-Tall (13)
Dayr al-Qasi (10)	al-Mansura (7)	Tarbikha (4)
al-Ghabisiyya (15)	Mi'ar (25)	Umm al-Faraj (12)
Iqrit (6)	al-Nabi Rubin (5)	al-Zeeb (9)
Iribbin Khirbat (2)	al-Nahr (14)	

It's an interactive map, and the dots are hyperlinks. Click on dot 23, and the server responds with census statistics and other facts about a community with 224 houses in 1931 and 1,460 inhabitants (1,330 Muslims and 130 Christians) in 1945. "Mostly destroyed" by the Israelis, al-Birwa once had a mosque, a church, and two schools. Although it's not apparent on the 1988 map, "three houses, two shrines, and one of the village schools remain standing." According to Palestinian historian Walid Khalidi, who is quoted extensively on the Web site, "All of these landmarks stand deserted amid cactuses, weeds, and fig, olive, and mulberry trees. The debris of destroyed houses punctuates the vegetation. There are also some graves near the site that are in a state of neglect. Part of the site and the land are farmed by the residents of Achihud," labeled Ahihud on the 1988 map. The second of two "Israeli settlements on town lands" is Kibbutz Yas'ur, just across Route 70. Hyperlinks summon several photographs, one showing "the houses of the Jewish settler [sic] built on stolen lands." Bitterness seems unavoidable but restrained.

PalestineRemembered.com is an electronic version of what Meron Benvenisti calls the "Arab map of Israel," a motley collection of mostly small-scale maps designed "to perpetuate the names of Arab villages that had been destroyed."[47] Web sites bring another dimension to the Palestinian complaint with charges of ethnic cleansing and demands for reparations if not repatriation for 700,000 Arab refugees.[48] Deir Yassin Remembered (www.deiryassin.org) recalls April 8, 1948, when "the mainstream Jewish defense force, the Haganah, authorized [an attack by] irregular terrorist forces of the Irgun and the Stern Gang," and "over 100 men, women, and children were systematically murdered."[49] Although repatriation is improbable—weekly attacks on civilians by suicide bombers only stiffen Israeli resistance to increasing its Arab minority—refugees have lingering memories of hastily abandoned property valued at over $24 billion.[50] The United Nations Conciliation Commission for Palestine, formed in 1948, has yet to negotiate a settlement, but its massive database includes a computerized geographic information system linking property records to scanned images of 5,625 maps.[51] Place names are surely a key component of the data, which remain confidential.

As the maps of Cyprus and Israel illustrate, toponymy acquires a special significance when ethnic groups with different languages covet the same territory. Plastered across a country's maps, place names assert ownership, legitimize conquest, and flaunt control. To the victor goes the toponymy along with other spoils of war. But as Palestinian Web sites demonstrate, the losing side can make its own maps, designed to refresh memory, sustain dreams, and reinforce resentment. Essential for identifying places, geographic names possess a symbolic power that can inflame as well as claim.

CHAPTER EIGHT

Inscriptions

With a small staff and a daunting mission, the U.S. Board on Geographic Names has little interest in policing the names of city streets, housing developments, and commercial centers, rarely labeled on Geological Survey topographic maps. An order of magnitude more numerous than natural features, local roads and shopping malls are best left to municipal governments and private landowners. Equally peripheral are politically ambiguous places like Antarctica, the ocean floor, and outer space, where effective standardization calls for cooperation with neighboring countries, the United Nations, and recognized scientific organizations like the International Hydrographic Organization and the International Astronomical Union. This chapter looks at naming controversies beneath or beyond the federal board's normal jurisdiction.

In the hands of local officials, street naming—and more significantly, renaming—is prone to pressures and complaints radically different from the challenges confronting federal and state names authorities. Not constrained by national policy on duplication and commemorative naming, local lawmakers can tack secondary names onto one- or two-block stretches or rename streets for living or recently deceased persons never connected with the city or neighborhood. Because reelection often depends upon prompt response to local concerns, politicians with vocal constituencies lose little time in commemorating military units and local heroes, replacing offensive or derogatory terms like *squaw*, or honoring slain leaders like John F. Kennedy and Martin Luther King Jr.[1] In municipal toponymy this capacity for change makes change inevitable.

Restrictions are few and largely reflect commonsense concerns that two streets with the same name might confuse firefighters or ambulance drivers at a time when minutes matter. Because Emergency 911 dispatching is usually a metropolitan or countywide service, city officials eager to rename a street typically need regional approval. If there is any likelihood that adjoining municipalities might someday merge, planning officials are justifiably leery when two streets that don't join seamlessly share a name. Quebec City, in Canada, learned the hard way after a merger of thirteen municipalities in 2002 required the renaming of 922 streets.[2] Renaming is always an annoyance for residents, a hassle for letter carriers and mapmakers, and a significant loss for business owners with thousands of dollars invested in stationery and advertising.

City officials wary of forcing address changes on constituents occasionally overlay a commemorative name on an otherwise unaltered street map. Thus an oversize traffic island becomes a "plaza," an ordinary intersection morphs into a "square," and a one- or two-block stretch of a street that retains its original name is dubbed a "boulevard" or "parkway." Syracuse, New York, took the latter route a few years ago when a veteran broadcaster retired and a two-block portion of James Street near the television station where he had worked became Ron Curtis Parkway. Although a county law passed in 1972 required formal approval by the county planning board, the mayor autocratically ordered the Department of Public Works to install white-on-blue parkway signs beneath the white-on-green markers for James Street. When the county planning director complained that the new signage "creates the potential for confusion," city officials argued that the renaming was merely "a ceremonial thing . . . with no legal impact."[3] I don't know how the issue was resolved, but the blue signs are still up.

Public-safety objections proved pivotal in Auburn, New York, a small city thirty miles west of Syracuse. Auburn was the home of William Henry Seward, the secretary of state who negotiated the

Alaskan purchase, and Harriet Tubman, a former slave who helped run the Underground Railroad. A courthouse plaque and several historical markers commemorate her activism, and the Harriet Tubman Home is an important local landmark, albeit less impressive then the Seward Mansion. Recent attempts to honor Tubman by renaming an elementary school, the high school auditorium, and a portion of U.S. 20 known as the Arterial encountered an embarrassing resistance that reflects local conservatism if not strained race relations.[4] Parents and alums didn't want Tubman's name on their school, longtime residents were comfortable with the name Arterial, and a retired fire captain on the five-member city council convinced two colleagues that Harriet Tubman Memorial Parkway might easily be confused with Tubman Lane, an isolated block-long street not shown on some maps.[5] Undaunted, leaders of the remember-Tubman movement persuaded the school board to designate their headquarters the Harriet Tubman Administration Building.

Racial and ethnic naming, renaming, and overlay naming is a fact of life in American cities with influential minority populations easily persuaded that squares and plazas provide suitable naming opportunities after all the parks and playgrounds are taken. Between 1978 and 1983, for instance, the New York City Council appended labels like Pope John Paul II Square and W. C. Handy Place to 105 venues throughout the five boroughs.[6] City officials must be wary that honoring one group can trigger complaints from a competitor. In 1986, when a council member proposed naming a portion of East Forty-third Street to honor David Ben-Gurion, Israel's first prime minister, Ann Leggett, speaking for the American-Arab Anti-Discrimination Committee, suggested adding balance by naming a street for Yasir Arafat.[7] Ten years earlier the council tabled a proposal to change Graham Avenue in Brooklyn to Avenue of Puerto Rico after longtime Italian residents of the neighborhood showed up with an Italian flag and confronted the measure's sponsor with hostile questions about the relative contributions of Italian Americans and Puerto Ricans.[8]

Renaming can be deliberately confrontational, as in December 1984, when New York mayor Edward Koch dedicated the street corner opposite South Africa's UN mission to Nelson and Winnie Mandela.[9] Koch ordered a sign put up after the proposal had been stalled in the city council's parks committee for five months. His action got committee members' attention. They endorsed the measure when the mayor agreed to take down the sign if the full council rejected it at their next meeting. "But that," he told the *New York Times,* "is not going to happen."

Developers eager to attract middle- and upper-class white Americans prefer a more genteel pitch free of ethnic overtones. As cultural geographer Wilbur Zelinsky observed, their focus on image-making yields a numbing similarity "relying on real or synthetic place names and terms imported from England, Scotland, and Wales (but not Ireland), and increasingly from France, Spain, and Italy [but] almost never [from] Slavic or Scandinavian countries, Africa, the Middle East, or Asia."[10] A preference for wordy, poetic allusions to rural landscapes and terms suggesting community is apparent in names like the Clusters of Meadowview (a subdivision near Ypsilanti, Michigan), Lawn Haven Burial Estates (a cemetery northeast of Pittsburgh), and Marketplace at the Grove (a shopping center near San Diego). Besides bemoaning developers' aversion to accurate description, Zelinsky accused them of "turning the country into a place where you can be both everywhere and nowhere."[11]

Recall the three principles of real estate—location, location, and location? For potential home buyers scanning classified ads, neighborhood names are far more significant than latitude and longitude. And because cachet and cash are synonymous, sellers often latch onto a more upscale area next door with toponymic gambits like Greenwood East, which is easier to do in older neighborhoods where subdivision boundaries are not inscribed in networks of culs-de-sac. A recent example cited by social critic Gregory Rodriguez in *Los Angeles* magazine is prosperous Hancock Park, an L.A. neighborhood that "managed to swallow up" smaller adjoining areas once known as Van Ness Square and Wilshire Highland Square.[12] As Rodriguez noted, renaming is a convenient ploy for residents dissatisfied less with their neighborhood than with its name. Because of the federal board's hands-off policy toward subdivisions and other features not considered "natural," a home owners association or local chamber of commerce can select a new name and ruthlessly promote a makeover with maps, signs, and strategic landscaping. Support of local government is helpful, though, if you want quick and certain acceptance of your new identity.

Toponymic makeovers come in two varieties: renovation, if only the name is the problem, and secession, when a newer or newly gentrified neighborhood seeks to distance itself from the larger, less at-

tractive place it lies within. Journalist Scott Shuger, who called attention to neighborhood renaming in Los Angeles in the early 1990s,
highlighted the secession of affluent West Hills from Canoga Park
in 1987.[13] Initiated by realtors, supported by the West Hills Chamber
of Commerce, and endorsed by the L.A. City Council, the divorce
boosted property values in West Hills by 15 percent. The importance of
a fresher, more bucolic name was equally apparent when North Hills
separated from Sepulveda, Lake View Terrace pulled out of Pacoima,
and Valley Village severed ties with North Hollywood. Among several
instances of simple renaming was the substitution of the Indian
name Canoga Park for Owensmouth, named for the Owens River, in
the early 1930s. According to Shuger, a local women's club initiated the
change to remove "any subsequent taint" of the scandalous diver-
sion of water from the Owens River, dramatized decades later in the
Roman Polanski film *Chinatown*.[14]

Although the Board on Geographic Names normally does not rule
on names of subdivisions and shopping centers, the Geographic
Names Information System records toponyms for churches, schools,
cemeteries, hospitals, and other cultural features beyond the federal
board's purview yet nonetheless useful to federal agencies, and thus
"official" once added to GNIS.[15] A subdivision, shopping center, or
similar private development would typically be listed as either a populated place or a locale, the latter defined as a "place at which there is or
was human activity [but] not includ[ing] populated places, mines, and
dams."[16] And when necessary, the federal board will choose between
competing names or renderings, which might arise, as its *Principles,
Policies, and Procedures* manual notes, "when several property owners
are involved." What's more, the FAQ (frequently asked questions) section of the GNIS Web site invites citizens to submit names not yet in
the system.

> 7. *Can I add new entries for manmade and administrative features, such as
> churches, cemeteries, schools, shopping centers, etc.?*

> Yes, for names that are not natural features, simply submit to us at
> this Web site the official name of the feature, its precise location, and
> a bibliographic reference, that is, a reference to a written source such
> as a map, pamphlet, other document, Web site, sign, etc. on which the
> name is published. Upon verification, the name will be entered into

the GNIS database. If a precise location is not available or submitted, the geographical coordinates will be entered as "unknown."[17]

As the answer implies, up-to-date coverage of "manmade" places can be problematic in areas with rapid suburban growth and in states for which Phase II names compilation (see page 31) is incomplete. For example, I found only twelve shopping centers or malls listed for New York State, where Phase II is still a vague promise, while for Nevada, with Phase II completed in 1991, I identified at least fifty-three retail centers in Clark County (Las Vegas) alone.[18] Although privately owned locales are valuable landmarks for motorists, keeping the list current is a continuing struggle, even with the assistance of volunteer informants.

One type of name we'll never see in GNIS, not even among the roads and streets planned for Phase III (or whatever supercedes it), is the "trap" feature once used by map publishers to protect their copyrights. Wary of competitors who would exploit their laborious field-checking and painstaking compilation of grid-referenced street indexes, mapping firms would insert a few fictitious streets, replete with plausible names, to incriminate a sloppy plagiarizer. With no ready defense against blatant infringement, the lazy rival would either pay a hefty settlement or file for bankruptcy. This game of gotcha became pointless in the mid-1990s after the U.S. Supreme Court ruled in *Feist v. Rural Telephone Company* that the Copyright Act did not protect facts, and a federal district court ruled in *Alexandria Drafting Co. v. Amsterdam* that false facts like trap streets were facts, nonetheless, and thus not eligible for copyright protection.[19]

It's difficult to find trap streets because cartographic publishers revealed them only in plaintiff's briefs, but I think I found one in my neighborhood. Figure 8.1 shows part of a Syracuse suburb as shown on maps published in 1979 and 1997 by Marshall Penn-York, a local mapping firm that apparently inserted "Gould Street" as a trap on its 1979 edition. At least I've uncovered no evidence the block-long street was ever built or even planned. But as the firm's updated, redesigned pre-millennial edition indicates, it's gone now, thanks (I assume) to *Feist*.

*   *   *

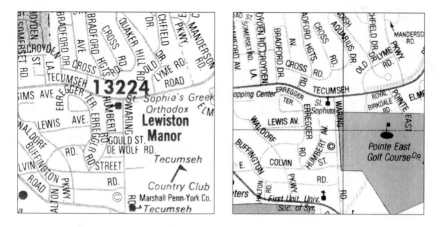

Figure 8.1. Gould Street, just below the center of the excerpt from a 1979 street map (*left*) is probably a trap street, inserted to detect copyright infringement. It was left off a revised edition issued in 1997 (*right*), after the federal courts made trap features worthless. Excerpts from Marshall Penn-York Co., *Visual Encyclopedia® Fully Indexed Street Map of Metropolitan Syracuse and Onondaga County, New York,* and Marshall Penn-York Co., *Visual Encyclopedia® Map of Syracuse, N.Y.*

Trap streets are no less whimsical than the boundary lines superimposed on Antarctica in the early twentieth century by the compulsive colonizers Britain and France, joined by Argentina, Australia, Chile, and New Zealand, which carved out sectors of territorial sovereignty with meridians extending poleward from the 60th parallel. Justifications included discovery, exploration, and proximity.[20] A seventh claimant, Norway, asserted rights based on mapping as well as prior exploration.[21] Although permanent settlement was never likely, the claims promised increased cartographic prominence as well as perpetual mining rights.

An Australian government Web site intended for educators describes the claims cartographically in a sample lesson plan on "Sovereignty and Territorial Claims" in Antarctica.[22] Its map (fig. 8.2) shows overlapping Argentinean, British, and Chilean sectors diametrically opposite the Down Under nation's humungous slices of the Antarctic pie, separated ironically by a narrow, bladelike intrusion claimed by France. "A strong presence" here is important, teachers and students are told, because it "minimizes the risk of external threats to Australia from that direction."[23] Hardly the first preemptive land grab in the name of national defense.

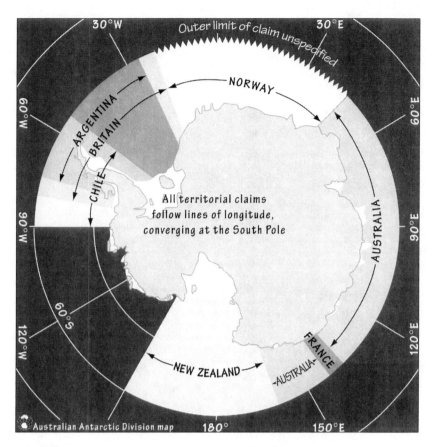

Figure 8.2. Australian government's generalized map of territorial claims for teachers. From Australian Antarctic Division, Classroom Antarctica, "International Sovereignty, Unit 7.1."

Territorial claims here have been frozen (no pun) since 1959, when the seven claimants and five other nations (including the United States and the Soviet Union) signed the Antarctic Treaty, which not only suspended all claims for the pact's duration but also encouraged scientific study and banned weapons testing, military maneuvers, and mineral exploration.[24] The United States initiated the treaty to preserve the scientific cooperation that culminated in the International Geophysical Year (1957–58). Russia and the United States had explored Antarctica off and on since the 1820s, and neither country recognized any of the claims. Refusing to let artificial borders obstruct their scientific endeavors, the two postwar superpowers established research stations, conducted aerial surveys, and named features wherever they pleased. And they're far from alone: according to the Central Intelligence Agency's map of the continent, the roster of nonclaimant nations with year-round research stations now includes China, Germany, India, Italy, Japan, South Africa, and Ukraine.[25] Although the CIA map describes the claims, the presence of one country's research base inside another's territorial boundary (fig. 8.3) exposes Antarctic sovereignty as little more than geopolitical posturing.

Rhetorical boundaries are not the only way to assert property rights. Mapping and naming declares an interest that might develop later into a formal, more traditionally delineated claim. Applied toponymy reinforced the American presence in Antarctica from the mid-1930s onward, when the United States encouraged its explorers and scientists to plant the flag across the continent.[26] This departure from Washington's earlier policy of merely refusing to acknowledge other countries' claims no doubt reflects anxiety over accelerated boundary drawing—four of the seven pie-slice claims were announced between 1933 and

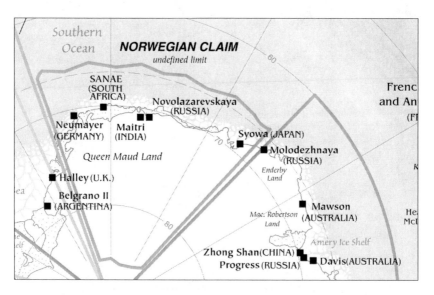

Figure 8.3. As shown on this excerpt from a CIA map of Antarctica, Norway's claim as well as the westernmost portion of the Australian claim (*lower right*) include year-round research stations set up by six nonclaimant nations. Below the excerpt at the South Pole is the Amundsen-Scott station operated by the United States. Enlarged for clarity to approximately 1:37,500,000 from U.S. Central Intelligence Agency, "Antarctic Region."

1942. Photogrammetric techniques pioneered in the 1930s and per-
fected during World War II fostered detailed mapping of the coastal
fringe at scales of 1:250,000 or greater, which revealed numerous
hills, cliffs, glaciers, capes, coves, and other natural features, all ripe
for toponymic recognition.[27] What better way to promote the American presence than a continent-wide gazetteer honoring hundreds of
scientists, engineers, and other contributors, mostly American? Later
on, after the treaty was signed, Yankee gazetteers were equally adept in
promoting Antarctica as a scientific commons.

America's first Antarctic gazetteer appeared in 1947, as a special
publication of the Board on Geographic Names.[28] Initiated four years
earlier by the board's Special Committee on Antarctic Names (later renamed the Advisory Committee on Antarctic Names, or ACAN), the
gazetteer included a 92-page list of expeditions, a 16-page bibliography, and a 127-page list of roughly 1,400 names for about 800 unique
features. Charged with standardizing toponyms for use by the federal
government, the committee examined maps and lists of names in several languages, researched each feature's historical background to resolve discrepancies in spelling or application, listed variants as well as
approved names, and translated a number of nonpersonal names
(mostly Scandinavian and German) into English. Research revealed
numerous superimposed names attributed to either an imprecise
description of location or deliberate duplication by a claimant nation
eager to supplant a rival's toponyms. Addressing the needs of American users, the committee imposed an anglicized nomenclature of feature types. A second edition, published in 1966 with 8,500 decision
entries and 3,000 variants, reflected a shift from historical research
to systematic screening of new names proposed by Americans and
others.[29] An updated listing issued in 1981 contained 12,000 approved
names,[30] a 1995 edition added 700 more,[31] and the online version at the
GNIS Web site provides access to nearly 14,000 official names, along
with more than 4,400 variants.[32]

Maurice Aurousseau, the prominent British names expert who
evaluated the 1947 gazetteer for the American Geographical Society's
Geographical Review, acknowledged the committee's "careful sorting
and sifting," applauded its "fair-mindedness and impartiality," and
pronounced the result "a volume of scholarly comment which no student of Antarctica can afford to neglect."[33] Even so, he questioned both
the slavish romanization of existing names and the heavy-handed

shortening of long names like "King George the Third's Sound," reduced to "George the Third Sound" by rules eliminating titles like *king* and *kaiser,* and thus hidden in an alphabetized list from anyone looking for the original name. Even more troubling was the elimination of duplication by transferring proper names to a different type of feature, previously unnamed. For example, if one explorer named a mountain after Jack while another named it after Jill, the committee apparently apologized for making Mount Jill a variant by adding a Jill Cliff or a Jill Glacier to the approved list. Aurousseau was particularly peeved that the United States, obsessively pursuing standardization, would tinker with names conferred by explorers from other nations. "Foreign geographical names . . . in uninhabited country [deserve] our respect," he argued. "On their irregularity, indeed on their irrationality, depends much of their effectiveness."[34]

Much perhaps, but how much? Federal policy on Antarctic names recognizes the tension between an explorer's naming rights and the toponymist's need for consistency, and it's not totally rigid. Despite a mandate to change or drop names deemed inappropriate because of duplication, structural flaws, or irrelevance to Antarctic exploration, ACAN guidelines promise due respect for "wideness of acceptance, as evidenced by extended use on maps and in the literature."[35] Names otherwise banned include those commemorating friends, relatives, products, sled dogs, pets, and donors "who by the nature and tone of their advertising have endeavored to capitalize or to gain some commercial advantage as a result of their donations."[36]

Wary of naming a prominent feature for a porter or technician, ACAN sorted features and potential honorees into three categories. Names of first-order features like coasts, mountain ranges, and large glaciers are reserved for individuals who have led, organized, or sponsored expeditions or for explorers or scientists who made important discoveries. By contrast, third-order features like cliffs, hills, and coves can commemorate expedition members, facilitators, supporters, and even "teachers or administrators in institutions of higher learning who have contributed to the training of polar explorers."[37] Among the Antarctic features described in GNIS is Barlow Rocks, named for U.S. Geological Survey cartographer Roger Barlow. A former student of mine, Roger spent a year at the South Pole Station in 1992, as a member of a satellite surveying team. In marked contrast to the Domestic Names Committee's ban on names commemorating living persons or the recently deceased, ACAN can honor anyone with a clear connection to the continent.

In addition to its "one feature per person" rule, which avoids confusion and rations naming opportunities, ACAN resists naming different features of the same type for persons with the same surname. Exceptions are essential, though, particularly for honorees named Smith, commemorated not only by a bluff, a cape, a cliff, a glacier, a "heights," an island, a knob, a lake, a "mount," a nunatak (hill surrounded by glacial ice), a peak, a peninsula, a point, and a "rocks" but also by two inlets, two ridges, and four duplicative plurals (Smith Bluffs, Smith Islands, Smith Nunataks, and Smith Peaks) as well as the possessive sans apostrophe Smiths Bench, named for "William M. Smith, psychologist, a member of the USARP [Antarctic Research Program] Victoria Land Traverse Party which surveyed this area in 1959–60."[38] Each member of an Antarctic expedition had an important responsibility, and it's fitting to commemorate them all, diesel mechanics and nurses as well as geophysicists and surveyors.

Cooperation with names authorities in Australia, Britain, and New Zealand avoided much of the duplication inevitable when other nations active in Antarctica and eager to honor their own explorers and scientists set up names databases reflecting their individual alphabets, linguistic conventions, and standardization principles. Wider coordination followed initiation of the Composite Gazetteer of Antarctica in 1992 as a project of the SCAR Working Group on Geodesy and Geographic Information.[39] (SCAR is the Scientific Committee on Antarctic Research, an international organization affiliated with the Antarctic Treaty Organization and the International Council for Science.[40]) An online database, which supplements a paper edition published in 1998, now includes over 35,000 official names submitted by twenty-three countries and representing 17,500 unique features, each identified by a feature number and linked to one or more approved names. For example, reference number 8206 identifies an island within the Argentinean, British, and Chilean claim lines that Great Britain, Russia, and the United States call Lavoisier Island, Argentina labels Isla Mitre, and Chile recognizes as Isla Serrano. Approximately 60 percent of features in the database were named by more than one country, according to researchers at the Italian National Agency for New Technology, Energy and the Environment (ENEA), who compiled the gazetteer using maps, geographic coordinates, and feature-class codes submitted by SCAR members. Although transliteration was necessary for submissions from Bulgaria, China, Japan, Poland, and Russia, the compilers used multiple fonts to accommodate all diacritical marks. Because political sensitivity precluded an effort to correct errors or reconcile diverse criteria for inclusion, coverage is inherently uneven. Despite this unevenness, the Composite Gazetteer is a valuable tool for countries committed to SCAR's belief that "national committees [should] avoid adding new place names to features already named."[41]

*   *   *

Worldwide standardization also eludes the naming of undersea features, addressed separately by two key committees, one American and the other international. The American authority is the Advisory Committee on Undersea Features (ACUF), affiliated with the Foreign Names Committee of the U.S. Board on Geographic Names in much the same way that ACAN works with the Domestic Names Committee.[42] While ACAN's electronic gazetteer is a distinct component of the GNIS Web site, ACUF's decisions are interspersed with other official foreign names on the GEOnet Names Server (GNS), maintained by the National Geospatial-Intelligence Agency.[43] The international body is the GEBCO Sub-Committee on Undersea Feature Names (SCUFN). GEBCO is the General Bathymetric Chart of the Oceans, a joint project of the International Hydrographic Organization (IHO) and the Intergovernmental Oceanographic Commission of the United Nations Educational, Scientific, and Cultural Organization (UNESCO). Fortunately for readers annoyed by acronyms, this alphabet soup has only two key ingredients: ACUF and SCUFN.

America's independence in naming undersea features partly reflects innovative development of bathymetric techniques in the 1950s and 1960s, including the precision depth recorder, an improved echo-sounding system that helped marine cartographer Marie Tharp and geophysicist Bruce Heezen map the Mid-Atlantic Ridge in sufficient detail to identify seafloor spreading as a mechanism of continental drift, a theory largely ridiculed by geologists until the mid-1960s.[44] Their maps revealed a deep rift valley running down the center of a 10,000-mile-long ridge that marked the divergence of the North American and Eurasian plates. In addition to confirming a crucial hypothesis in modern geophysics, detailed maps of the ocean floor exposed numerous hitherto unknown undersea features as targets for commemorative naming. In 1963 the U.S. Board on Geographic Names established the Advisory Committee on Undersea Features to coordinate toponymy for a growing number of federal bathymetric charts. ACUF issued four gazetteers between 1961 and 1990, the last with roughly 10,000 names.[45]

SCUFN's ancestry traces back to 1903, when Prince Albert of Monaco offered to coordinate compilation of a 24-sheet map of the oceans based on 18,000 soundings, mostly from the British Admiralty.[46] Published at an equatorial scale of 1:10,000,000, the General Bathymetric Chart of the Oceans identified a few important undersea features by name,[47] and subsequent editions initiated in 1910, 1929, and 1954 steadily increased the roster of named features. The International Hydrographic Organization, which coordinated the third and fourth editions, established a Geographical Names Sub-Committee, later renamed the Sub-Committee on Undersea Feature Names. In 1973 GEBCO became a joint venture of the IHO and the Intergovernmental Oceanographic Commission (IOC), a UNESCO affiliate eager to incorporate better data as well as make the chart's contour lines more accurate.[48] A fifth edition, initiated in 1974, led to the GEBCO Digital Atlas, published on CD-ROM in 1994 and followed by an updated "Centenary" edition in 2003. An electronic gazetteer, regularly revised and available at a Web site hosted by the National Oceanic and Atmospheric Administration, currently contains over 3,400 names, including fracture zones named for Tharp and Heezen.[49] SCUFN's list is inherently shorter than ACUF's because the 1:10,000,000 GEBCO is less detailed than most American bathymetric charts. Growth of the SCUFN roster also reflects the IOC's International Bathymetric Charts, available at 1:1,000,000 for the Mediterranean Sea, the Central Eastern Atlantic, and a few other regions. Like other international organizations active in toponymy, the IHO and the IOC offer guidelines but do not adjudicate disputes or set international policy.

Early efforts to standardize undersea toponymy focused on nomenclature. In 1954, when deep soundings were comparatively sparse, the International Committee on the Nomenclature of Ocean Bottom Features (a SCUFN predecessor) authorized only five generic names for clearly isolated features, with relief differentiating the tall, cone-shaped *seamount* and flat-topped *tablemount* from the shorter *seapeak* or *seaknoll* and the broad, comparatively flat *oceanic bank*.[50] Where isolation of individual features was uncertain, a significant protrusion from the ocean floor was called a *seahigh*. By 1967 GEBCO had developed terms and definitions for forty-four different features ranging in size from a *continental shelf* to a *gully* and defined by geometry rather than plausible origin.[51] Although any seamount is most likely an underwater volcano—a 1,000-meter (3,300-foot) rise from the ocean floor is required—the hydrographic toponymist leaves interpretation to the marine geologist. Despite four decades of editorial refinement, SCUFN's current list includes forty-four largely similar feature types as well as cross-listings for nine synonyms or obsolete terms.[52]

Proposals must include maps, photographs, or measurements demonstrating an accurate application of feature definitions. When the generic portion of a proposed name seems inappropriate, SCUFN insists on additional information or substitutes its preferred term. For instance, in 2003, when Germany's Alfred Wegener Institute for Polar and Marine Research proposed names for several previously unlabeled features in the Fram Strait, east of Greenland, SCUFN objected to the vague, unapproved term *deep* and changed the suggested names Molloy Deep and Spitsbergen Deep to Molloy Hole and Spitsbergen Trough.[53] To inform others about the importance of exact terminology and advertise SCUFN's acceptance of its recommendations, with modifications, the institute publicized the decision on its Web site with a vivid three-dimensional map differentiating a hole's circularity from a trough's pronounced elongation (fig. 8.4).

The Wegener Institute's choice of specific names posed no problems. Spitsbergen Trough satisfied guideline 3, whereby "the first choice of a specific term, where feasible, should be one associated with a geographical feature," while Molloy Hole, commemorating pioneering U.S. Navy research scientist Arthur E. Molloy, accords with either guideline 4—which permits names commemorating ships, vehicles, expeditions, or scientific institutes "involved in the discovering and/ or delineation of the feature, or to honor the memory of famous persons"—or guideline 5, which allows naming features after living persons "who have made an outstanding or fundamental contribution to the ocean sciences."[54]

While the international committee had no misgivings about honoring Molloy, SCUFN occasionally complains of ACUF's reluctance to question an American honoree's oceanographic credentials. Particularly troubling is the "commemoration of retiring US agency officials

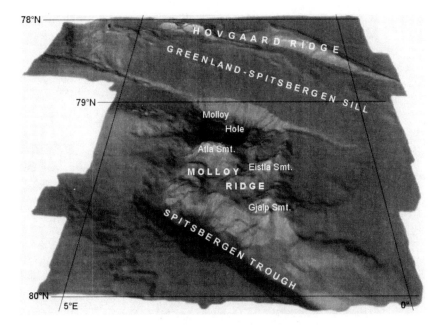

Figure 8.4. Oblique view of the Fram Strait illustrates SCUFN's substitution of *hole* and *trough* for the inappropriate generic *deep* in the original proposals for Molloy Hole and Spitsbergen Trough. "Smt." is an abbreviation for seamount. From Alfred Wegener Institute for Polar and Marine Research, "Fram Strait Feature Names."

and senior personnel."[55] SCUFN's 2002 annual report cites as an example Bernard Seamount, named for Landry J. Bernard, identified merely as a "civilian scientist at the US Naval Oceanographic Office."[56] (Bernard was technical director of the Naval Oceanographic Office for ten years, but you won't find him in *American Men and Women of Science*.[57]) "For appropriate international acceptance and lasting use," the report warns, "living individuals so commemorated should be known widely outside their own organizations."[58] Despite concerns that "naming a seamount for an individual has become almost a 're-tirement benefit,'" SCUFN routinely endorses ACUF's labeling of previously unnamed features. Its protests consist mostly of timid caveats like the plea that ACUF "consider the result [if] its current practices be given worldwide application by similar bodies."[59] Paradoxically, the two groups have identical policies on commemorating living persons.[60]

SCUFN and ACUF also disagree over terminology, notably the latter's "recent re-establishment . . . of the archaic and very subjective term 'deep' for variously localised depressions of subjectively determined scale in almost any depth regimen."[61] Although SCUFN objects to using *deep* in new names, it concedes the historical value of well-established names like Challenger Deep.[62]

Procedural disputes like these reflect disagreement over the worthiness of some honorees as well as a profound, perhaps unavoidable conflict between whiny advocates of international collaboration and toponymic imperialists eager to claim their country's discoveries. According to Richard Randall, executive secretary of the Foreign Names Committee from 1973 to 1993, SCUFN would like to be the sole authority for naming undersea features in international waters, while the federal board "sees its programs as fully able to meet the requirements of the broadest possible user group."[63] And as Trent Palmer, ACUF's current secretary, reminded me, SCUFN was established to service a series of small-scale ocean charts and "has no mandate to be an international authority on undersea feature names."[64] Despite the impasse, SCUFN and ACUF exchange lists and endorse most of each other's decisions, as in 2003, when the GEBCO subcommittee recognized the Randall Seamounts, honoring Richard, his father, and his two brothers, a quartet of federal cartographers and lexicographers. Acquiescent as ever, SCUFN reported heights and positions for the four seamounts but "left it to ACUF to take the lead in assigning individual names, should they so wish."[65]

*   *   *

Anyone worried about shrinking opportunities for commemorative naming need only look upward—the heavens encompass billions and billions of stars (as the late Carl Sagan is reputed to have said[66]) while our sun's planets and their satellites promise an abundance of identifiable surface features, many only recently visible and most of these not yet named. As Antarctica and the ocean floor illustrate, features in uninhabitable territory are poised for naming once they're mapped. And because multiple cartographers can create confusion, there's a pretext for applied toponymy, with its comprehensive gazetteers, systematic guidelines, and personal accolades, fixed and sanctioned for official use. For toponymists and their honorees, planetary cartography with its authenticated, viewable juxtaposition of names and features suffices for reality.

Unlike Antarctica or the ocean floor, outer space has a universally recognized naming authority, the International Astronomical Union.[67] Founded in 1919 to promote cooperative research, the IAU assumed responsibility for standardizing planetary toponymy, initiated around 1600 when William Gilbert embellished his naked-eye sketch of lunar seas and islands with thirteen place names.[68] The number of named features jumped to 325 in 1645, when Michael van Langren published an impressively detailed lunar chart based on telescopic observation.[69] Improved optics led to ever more profusely labeled lunar charts based on competing systems of lunar nomenclature as well as ambitious small-scale cartographic renderings of Mars, Venus, and the moons of Jupiter.

As Richard Proctor's late nineteenth-century map of Mars (fig. 8.5) illustrates, planetary cartographers conditioned by terrestrial globes and wall maps eagerly equated tonal differences with shorelines. Proctor based his chart on telescopic drawings by William Dawes and gave each feature two names, a specific name commemorating a fellow astronomer followed by a generic name evoking a similarly configured feature on Earth.[70] The first systematic nomenclature for the Red Planet,[71] Proctor's scheme was abandoned because of its imprecision and wasteful duplications like the continent, ocean, and sea named for Dawes.[72]

Post–World War II advances in rocketry and imaging technology yielded thousands of unprecedented planetary snapshots, which sci-

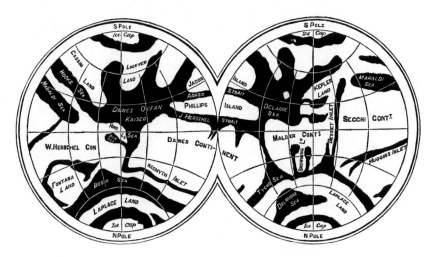

Figure 8.5. Richard Proctor's Martian chart, ca. 1868. From Proctor, *Other Worlds than Ours*, 109.

entists skilled in data management, image processing, map projection, and cartographic illustration prepared for public viewing by adding meridians, parallels, and feature labels.[73] Exceptionally detailed imaging of the lunar farside, Mars, and Venus in the 1960s and 1970s led to multi-sheet map series giving politicians something to show for their enormous investment in space exploration. Planetary atlases and gazetteers soon followed.[74] Figure 8.6, a photomap for a portion of the Martian surface about the size of North Dakota, attests to the intriguing realism and abundant naming opportunities of planetary cartography.[75]

A close look at the chart excerpt reveals craters named Edam, Kasimov, Murgoo, and Seminole for places in the Netherlands, Russia, Australia, and Florida, respectively. Although the very largest craters on Mars commemorate deceased scientists and science fiction writers who promoted the planet, most are named for villages and small cities on Earth. A second IAU-endorsed theme is apparent in the elongated depressions, or *valles* (plural of *vallis*, Latin for "valley"), labeled Clota, Oltis, and Samara, which carry the ancient names of the River Clyde (in Scotland) and the Lot and Somme rivers (both in France). According to IAU guidelines, the planet's largest *valles* bear the name for

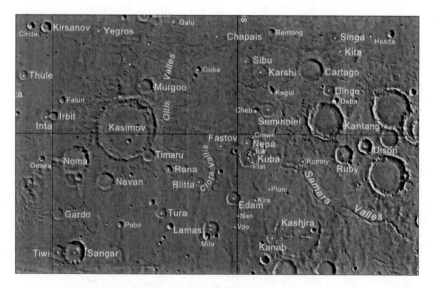

Figure 8.6. Excerpt from Mars 1:5 million–scale Viking imagery. Area shown is about the size of North Dakota. From U.S. Geological Survey, Astrogeology Research Program, "Gazetteer of Planetary Nomenclature," sheet MC-19.

Mars in various languages—examples include Al-Qahira from Arabic, Ares from Greek, and Auqakuh from Quechua, the language of the Incas—while most others are named for rivers on Earth.[76]

Other parts of the solar system have their own themes. Prominent craters on Venus, for instance, commemorate famous women, whereas craters on Mercury honor artists, musicians, poets, and writers from many cultures. Every mapped asteroid and satellite has its own themes.[77] On the asteroid Ida, for example, craters are named for caverns on Earth, and the two discernible craters on Ida's tiny moon, Dactyl, are named for two of the three original Dactyls (mythical Greek figures associated with Mount Ida, on Crete). And to the delight of anyone with a liberal arts background, on the asteroid Eros, craters with names like Bovary, Don Quixote, Heathcliff, and Lolita commemorate mythological and literary lovers.[78]

Underlying this refreshingly quirky approach to standardization is an eagerness to avoid duplication and resolve disputes. In the late 1960s, for instance, IAU officials brokered a settlement between the United States and the USSR over the Moon's farside—because a Russian space probe got there first, the Soviets felt entitled to name all readily discernible features for cosmonauts and other fellow countrymen.[79] Although the IAU will modify or move newly applied names to avoid confusion, it retains well-established names, even those at odds with the current theme. Guidelines designed to avoid resentment and mistrust prohibit names with military, religious, or strongly nationalist overtones as well as toponyms commemorating anyone deceased less than three years.[80] Committed to a nomenclature that is "international in its choice of names," the IAU recommends an "equitable selection of names from ethnic groups/countries on each map." Although the latter rule accounts for the geographic diversity of towns and small cities commemorated by Martian craters, "a higher percentage of names from the country planning a landing is allowed on landing site maps."[81]

Acceptance of IAU authority in planetary nomenclature is perhaps surprising in light of America's unilateralist tendencies and its Cold War rivalry with the Soviet Union. While the United States would no doubt act independently if the organization failed to meet its requirements,[82] a strong American presence on the IAU Working Group for Planetary System Nomenclature (WGPSN) and its various "task groups" (subcommittees) favors cooperation. According to the working group's Web site, hosted by the U.S. Geological Survey's Astrogeology Research Program, five of the thirteen members of the WGPSN steering committee are American, as are three of the six or seven members of the task groups focusing on the Moon, Mars, "small bodies," and the "outer solar system."[83] U.S. representation is numerically weakest on the three-member Venus subcommittee, which includes two Russians and one American, and strongest on the four-member Mercury subcommittee, which consists of three Americans and one Briton.

Naming is ostensibly open and collaborative. Anyone may submit a name to the appropriate task group, the first step in a three-stage process. It's not necessary to pinpoint a specific feature insofar as task groups can bank names that meet IAU guidelines and assign them when a scientist eager to describe a particular feature requests a toponym.[84] Each group forwards acceptable names and feature descriptions to the WGPSN, which reviews proposals and sends a list of approved names to the IAU General Assembly, which meets every third year. Although provisional names can be used, pending higher approval, the General Assembly has the ultimate say, and no name is official without its endorsement.

Outside our solar system, IAU preeminence faces the challenge of entrepreneurial marketers like the International Star Registry, which promises that "for $54, plus shipping and handling, you can name a star." Part of the price covers a "beautiful 12″ × 16″ parchment certificate . . . with the name of your choice, dedication date, and telescopic coordinates of the star [and] a larger, more detailed chart with the star you name encircled in red." Backing up the certificate and chart is a suggestion of official endorsement and immortality:

> Because these star names are copyrighted with their telescopic coordinates in the book, "Your Place in the Cosmos," future generations may identify the star name in the directory and, using a telescope, locate the actual star in the sky.[85]

(Be leery of the first thirteen words—the Copyright Office copyrights the book, not individual star names.[86] What's more, the book's copyright might even discourage others from using your chosen star name.) Your $54 also covers online marketing through www.starregistry.com as well as radio ads that run several weeks before Valentine's Day, when parchment certificates identifying their very own star no doubt impress thousands of impressionable sweet-hearts. Much of the $54 is pure profit and an inspiration for Web-based competitors like Name a Star, Inc., which for only $24.95 offers the convenience of printing the certificate at home if your sweetie won't mind a smaller format and plain paper.[87] Query Google about "star names," and you'll find a half-dozen "supporting links" to other firms offering budget packages as well as upscale framing, memorials for newborns or deceased loved ones, special formats for Mother's Day or Father's Day, and overnight shipping.

Surprisingly perhaps, the IAU's reaction is mild amusement, not scornful condemnation. "As an international scientific organization," its Web site notes, "the IAU dissociates itself entirely from the com-mercial practice of 'selling' fictitious star names or 'real estate' on other planets or moons in the Solar System." While "the 'gift' of a star may open someone's eyes to the beauty of the night sky . . . it does not justify deceiving people into believing that real star names can be bought like any other commodity." What's more, the IAU "Layman's Guide to Naming Stars" observes, real astronomers don't name stars. The organization maintains a catalog of celestial bodies outside our solar system, but identifies them by coordinates and catalog number, not by name, so that researchers don't need to consult a gazetteer (and the IAU doesn't have to cope with managing applied toponymy for "hundreds of millions of stars"). Although it refuses to list star sellers, the IAU counsels buyers that the names they purchase are likely to be unique, at least on each company's list, "otherwise you can probably sue them."[88] But because companies don't pool their lists, some stars surely acquire multiple names. In short, without certification by an of-ficial, internationally recognized names authority like the IAU, a star name's uniqueness and worth are questionable.

However established, international recognition is very much a po-litical construction, arising from a shared recognition that cooperative standardization is preferable to competitive chaos. Aside from obviat-ing the confusion of multiple names for the same feature, cooperative naming lessens the likelihood that unilateralist gazetteers imposed by ACAN and ACUF will someday give way to a revolutionary, revisionist nomenclature of a planetary government committed to a more inclu-sive toponymy or the glorification of some other culture. After all, the Roman, Spanish, and British empires eventually atrophied. Better for the long run to consolidate the spoils of America's cartographic productivity and scientific preeminence by settling for a lion's share of an international toponymy coordinated by SCAR, SCUFN, and the IAU. In opting for limited international cooperation and the assumed permanence of what could become well-established names, the United States seems at least subliminally aware that standardization does not confer durability. Better to name some of the features for all time than all of the features for a mere century or so.

How long does an official toponym last? Clearly not forever in the case of local street names, which can be recycled by a city council or wiped out by a bulldozer. (Baltimore, Maryland—where my father's ancestors found refuge after fleeing Haiti in the early nineteenth century—had a Monmonier Court until the 1950s, when it succumbed to urban renewal.) But long enough when applied to nunataks and seamounts to mollify most honorees and their survivors. Of course, nature can obliterate natural features—the specter of global warming makes glaciers and low-lying rocks especially vulnerable—and future civilizations could recycle naming rights in much the same way medieval congregations cleared overcrowded churchyards by putting disinterred bones in an ossuary. Even so, obsolete names could persist indefinitely as variants if electronic gazetteers prove as durable as other forms of recorded history. Thanks to mapping and applied toponymy, commemorative names can be a rare honor even when toponyms or features don't last.

Naming Rites

"What's in a name?" Shakespeare once asked. Quite a bit, actually, as the foregoing chapters illustrate. Toponyms define places as well as identify them and are worth knowing about as well as merely memorizing. Although place-name geography is an often-tedious prerequisite to cultural literacy, how a nation manipulates and preserves its place and feature names says a lot about its respect for history, minority rights, and indigenous culture. And while the value of geographic names might seem largely symbolic when a comparatively inaccessible undersea or extraterrestrial feature is named for a retiree or dead scientist, in other contexts naming and renaming can leverage the enormous power of toponyms used in daily speech, directions, news reports, postal addresses, and road signs.

It seems only natural that geographers and mapmakers control toponymy. What other than a gazetteer or reference map can store geographic names in a compact, conveniently accessible, and eminently durable format? And who else is as obsessively devoted to the systematic, orderly description of the landscape? By taking over where the explorer left off and pursuing a broader, more complete inventory of toponyms, the cartographer inherited the right to name the myriad lesser features that first arrivals were too busy to notice, much less describe. And for land acquired by conquest, rather than discovery, mapmakers helped generals decide what to keep, what to dismantle, and what merely to re-label. As readily manipulated scale models of a larger, more complicated reality, maps could secure the linguistic victory deemed necessary for settlement or assimilation.

Although mapmakers still ran the show, geographical naming took a more formal, systematic turn in the late nineteenth century, when central governments began to encourage their national mapping organizations to share responsibility for toponyms and gazetteers with a wider community of map users, mostly (if not exclusively) in other government departments. For the United States the pivotal year was 1890, when President Benjamin Harrison established the Board on Geographic Names. Canada took a similar step in 1897 by forming the Geographic Board of Canada. The American approach evolved to include a sharing of power and responsibility among federal, state, and tribal governments, while the Canadian model reflects the greater autonomy of provincial and First Nations governments. Great Britain took a different approach whereby the Ordnance Survey, the country's official mapmaker, controls names on its own products, and the Permanent Committee on Geographical Names for British Official Use, a semi-independent body established by the Royal Geographical Society in 1919, produces gazetteers covering other countries. Despite diverse approaches among the world's nations, applied toponymy evolved as a deliberative process typically orchestrated by interagency committees following official policies and formal procedures. Since 1967, when the first United Nations Conference on the Standardization of Geographical Names met in Geneva, numerous countries outside Europe and North America have codified procedures for naming features within and beyond their borders.

As examples throughout this book illustrate, responsibility for official names on federal maps gave the U.S. Board on Geographic Names an enormous influence on commercial cartography at home and abroad. Although only federal agencies are compelled to conform to the board's decisions, state and private mapmakers find it convenient to consult official sources like GNIS and its foreign-names counterpart, the GEOnet Names Server. As long as the federal government controls the country's basic maps, its decisions on place and feature names will guide other mapmakers, who by and large appreciate officially endorsed toponyms with standardized spellings.

A centralized names board proved especially useful when groups and individuals began to protest racial and ethnic insults inscribed on maps in less sensitive times. Although expunging pejorative names from maps has been a slow, often frustrating process, largely because names authorities insist upon suitable replacements, applied toponymy struck a balance between die-hard traditionalists committed to preserving socially obsolete paper landscapes and bureaucratic realists aware that geographic naming is a mediated process with responsibilities and consequences. An electronic gazetteer named GNIS (Geographic Names Information System) expedited this compromise by quickly and systematically identifying toponyms based on pejoratives like *squaw* while recording as variants topographic embarrassments like Montana's Squaw Nipple, recently renamed Deer Point but preserved indefinitely as indisputable evidence of the mapmaker's contribution, naive or otherwise, to the social construction of race. An inherently conservative process, applied toponymy extends the cartographic lives of questionable feature names like Florida's Jewfish Creek and Oregon's Whorehouse Meadow while suppressing duplication, resisting impulsive commemorative naming, and preserving the cartographic virginity of wilderness areas, where remaining nameless is very much a part of remaining wild.

It's understandable that marginalized groups are offended by derogatory toponyms. Maps are both a tool and an emblem of government, and their content implies official endorsement. Ironically perhaps, many if not most cartographic insults are a consequence of the mapmaker's trying to get it right by ferreting out local usage—part of what map historian Brian Harley called "collecting culture."[1] Although it's questionable in many cases whether the disparaging feature name was ever widely known or used, once inscribed on the map it becomes a cultural time bomb, hidden in plain sight and ready to trigger an explosive reaction when discovered decades later.

While names authorities are on solid ground when fixing place and feature names within their own borders, a government's need to map other parts of the world raises issues about the appropriate exercise of cartographic power to appease allies and penalize enemies. Should American mapmakers, official and commercial, acknowledge a foreign government's right to determine how we identify its cities, rivers, and other geographic features? What if the other country's alphabet is arcane, its writing system confusing, or its names difficult to pronounce? What if a country renames itself or its cities for ostensibly pointless political purposes? Or if the individuals dictating these changes seized power illegally, through a coup or invasion? Should mapmakers elsewhere comply, and thus appear to endorse a rogue regime? Should cartographic recognition depend upon diplomatic recognition? What if a renamed feature is well outside the revisionist's territorial waters, as when the Koreas sought to replace Sea of Japan with East Sea? While bureaucrats who control the cartographic rendering of foreign names on official maps are servants of foreign policy, private mapmakers and scholars are free to reject or accept politically inspired changes—or to apply parenthetical labels that help users cope with an apparent revision that might not prove permanent.

Labeling privileges are most powerful when a mapmaker replaces the toponyms of the vanquished with those of the victor. This substitution is particularly poignant when winners expel losers and resettle their lands, as in Northern Cyprus and Palestine, and stronger still when the map confirms destruction of the original settlements and purports to erase the refugee's right of return. However useful in reinforcing territorial acquisitions and recruiting new settlers, the map's authority is hardly absolute when displaced peoples can make maps that nourish their longing for the homeland and lust for revenge. It's questionable whether oppressive maps are more effective in the long run than their subversive counterparts.

Victors unwilling or unable to eject and wall off opponents face the dicey issue of minority toponyms, especially divisive in a bilingual or binational society. More problematic still are ill-conceived, autocratic efforts to tinker with a people's linguistic heritage, as in South Tyrol, annexed from Austria in 1919 by Italy, which imposed eight thousand Italian toponyms on a largely German-speaking population.[2] Ferjan Ormeling, a Dutch geographer who mapped the topographic suppression of linguistic minorities in Western Europe, found that resentment inflamed by linguistic scholars and ethnic politicians typically provokes a clumsy attempt to restore the original orthography.[3] Once geographic names have been codified in the dominant language, he observed, recovery of repressed minority toponyms can be a costly, error-prone undertaking. Instead of bilingual maps, often cluttered in densely settled areas and useful principally as an interim solution, Ormeling recommended a thoughtful effort to agree on an equitable language policy before any maps are revised.

Native Americans face a situation radically different from that of Western Europe's linguistic minorities. Although native resistance currently focuses on offensive words like *squaw,* resurgent interest in tribal languages could precipitate a restoration of indigenous toponyms, romanized for use on maps—because indigenous North American languages rely on oral tradition, not writing, there's no original orthography to restore. Immersion schools designed to increase the number of native speakers can rescue dying languages,[4] and with the aid of linguistics scholars, tribes without a written language can even invent one, based on the roman alphabet and appropriate diacritical marks.[5] However contrived, these improvised spellings merely underscore the notion of language as broadly symbolic and socially constructed. What's in a name? More than most of us realize.

Notes

CHAPTER ONE

1. This origin of Niggerhead Point is a matter of town folklore according to Marjorie Perez (Wayne County historian), in telephone discussion with author, July 30, 1992, and Shirley Eygnor (Huron town clerk), in telephone discussion with author, July 31, 1992.

2. Heinen, "Squaw."

3. For a concise etymological history of *squaw*, see Bright, "Sociolinguistics of the 'S-Word.'" Also see Goddard, "Since the Word Squaw Continues."

4. See, for example, *Phoenix New Times*, "Squaw Pique"; and Giago, "Walk Up to Any Indian Woman."

5. Okimoto, "Board Rejects 'Phoenix Peak'"; and Institute of American Indian Studies, "Arizona Pressured to Outlaw 'Squaw.'"

6. Bright, "Sociolinguistics of the 'S-Word.'"

7. Goldman, "Arizona's Squaw Peak."

8. Gehrke, "Piestewa Peak."

9. Ladwig, "When Cape Crusaders Played." Also see Morris, *Florida Place Names*, 30–31; and Randall, *Place Names*, 139.

10. Federal Writers' Project, *Florida*, 346.

11. See, for example, Stewart, "Classification"; and Stewart, *Names on the Globe*, esp. 86–88. For a concise summary of Stewart's contributions to American place naming, see Beeler, "George R. Stewart."

12. Vasiliev, "Naming of Moscows"; and Vasiliev, "Naming and Diffusion of Moscows."

13. I recall the name and its rationale from my childhood in Baltimore County, Maryland, but my recollection of one or more these surnames could be faulty.

14. Stewart, *Names on the Globe*, 13.

15. For an insightful discussion of applied aspects, see Payne, "Applied Toponymy."

16. U.S. Board on Geographic Names, "Domestic Geographic Name Proposal."

17. Roger Payne (executive secretary, U.S. Board on Geographic Names), in telephone discussion with author, May 30, 2003.

18. Additional policies, which codify miscellaneous practices, are part of a revised policy statement, under review.

19. Orth and Payne, *Principles, Policies, and Procedures*, (online edition), 14. Dates for the establishment of the one- and five-year waiting periods are from Payne, "The United States Board on Geographic Names."

20. Orth and Payne, *Principles, Policies, and Procedures* (online edition), 14.

21. Ibid., 13.

22. Bowley, "'Moose' on the Loose."

23. Orth and Payne, *Principles, Policies, and Procedures* (online edition), 16.

24. Ibid., 17.

25. Bower, "How'd Those Yellowstone Places Get Their Names?"

26. Elsasser, "Name That Place."

27. Payne, "United States Board on Geographic Names." These decisions also applied to noncartographic uses of these names, for example, in scientific reports.

28. If usage by a prominent African American law professor is any indication, *Negro* and *black* are still acceptable. See Kennedy, *Nigger*. In my own experience, *people of color* seems too "politically correct," if not mildly unctuous, while *black* is a conveniently concise term for most social scientists and the U.S. Census Bureau.

29. Variant names can be used in parentheses following the official name, but the policy now evolving is likely to include an explicit ban on the parenthetical use of variants containing *nigger, Jap*, and *squaw*.

30. U.S. Geological Survey, *United States Geological Survey*, 60.

31. Evans and others, *History*, 93.

32. U.S. Geological Survey, *United States Geological Survey*, 57; "names" are also mentioned on 61 and 63.

33. Gannett, *Manual*, 8, 9–10. An abridged version of the manual was reprinted in 1906 as USGS Bulletin 307.

34. U.S. Geological Survey, *United States Geological Survey*, 62.

35. Gannett, *Manual*, 1.

36. Except as noted, all quotations in this and the following two paragraphs are from Beaman, "Topographic Mapping," 226–27.

37. Birdseye, *Topographic Instructions*, 161–378, 329.

38. Ibid., 329.

39. All quotations in this paragraph are from ibid., 326–27.

40. Payne, "United States Board on Geographic Names," 179–82; and U.S. Board on Geographic Names, *First Report*, 6.

41. Payne and various reports of the board indicate that the restoration of the final *h* occurred in 1911, while another source reports the date as 1932. See Pizer, "Place Namers."

42. U.S. Board on Geographic Names, *Fourth Report*, 14–16.

43. Definition is from sec. 2(c) of the Wilderness Act of 1964 (P.L. 88–577). For the full text of the statute, see the public law library on Wilderness.Net.

44. Orth and Payne, *Principles, Policies, and Procedures* (online edition), policy IV, sec. 1.

45. For examples of rejected proposals, see Julyan, "Protecting the Endangered Blank Spots."

CHAPTER TWO

1. National Research Council, *Weaving a National Map*, 7–8.

2. For insights on the early adoption of copperplate engraving in cartography and the engraver's role in shaping the appearance of maps, see Verner, "Copperplate Engraving." For a concise, carefully researched history of copperplate engraving of topographic maps at the U.S. Geological Survey, see Phillips, "Copperplate Engraving." Several parts of the process were photomechanical. So that the engraver knew exactly where to cut the lines and letters, a technician photographed the topographer's manuscript map, drawn in black ink, and used the resulting negative to fix the image on a thin photosensitive emulsion spread over the copper plate. After the engraver cut the lines and letters into the copper, the photolithographer rubbed a waxy substance into the resulting grooves, wiped the rest of the surface clean, and transferred a waxy image of the map to the smooth, flat lithographic stone used to print the maps. For a concise history of photomechanical image transfer in cartography, see Cook, "Historical Role of Photomechanical Techniques." For abbreviated accounts of early map engraving and printing at the Geological Survey, see U.S. Geological Survey, *United States Geological Survey*, 113–17; and Beaman, "Topographic Mapping," esp. 336–37. The lithographic stone, which could accommodate a run of several hundred copies, thus saved the carefully engraved linework from unnecessary wear. Early in the twentieth century, the Geological Survey adopted offset lithography, a more efficient printing process in which a flexible metal plate mounted on a cylinder replaced the flat lithographic stone. Phillips, who thoroughly examined the Geological Survey's annual reports and other publications related to mapmaking, says only that offset lithography was "on the horizon by 1904 and probably soon used by the USGS." See Phillips, "Copperplate Engraving," 16.

3. For a flavor of cartographic experimentation with imagesetter technology, which came of age during the 1980s, see Allord and Hitt, "Linking Digital Technology to Printing Technology"; U.S. Geological Survey, "Automated Cartographic Lettering"; and Mattson, "Imagesetting."

4. Older USGS topographic maps in various map collections as well as the Geological Survey's *Topographic Instructions* indicate that green and red overprints were used selectively at least as early as the 1920s. See Beaman, "Topographic Mapping," 256–57, 317, 319, 344.

5. U.S. Geological Survey, *United States Geological Survey Annual Report, Fiscal Year 1975*, 43.

6. Between the phaseout of copper engraving and its replacement by plastic scribing, USGS cartographic technicians drafted color or feature separations in pen and ink on dimensionally stable "metal-mounted paper." See Evans and others, *History*, 272, 300. Scribing of negative images on plastic base materials evolved from the scribing of negative images on glass plates coated with an opaque emulsion. A related advance was the transition from lithographic printing on smooth, flat printing stones to photolithography using high-speed, multicolor rotary offset lithographic presses. See Koeman, "Application of Photography to Map Printing."

7. Robinson and Sale, *Elements of Cartography*, esp. 327.

8. See National Research Council, *Weaving a National Map*, 11, 24, 68. The National Map has eight layers or themes, one of which is geographic names. The other layers are digital orthorectified imagery, elevation, land cover, water features, transportation routes, boundaries, and buildings and other critical structures.

9. *Merriam-Webster's Geographical Dictionary*, 493, 782.

10. White, "National Gazetteers."

11. Haskel and Smith, *Complete Descriptive and Statistical Gazetteer*.

12. Steinwehr, *Centennial Gazetteer*.

13. White, "National Gazetteers," 15.

14. Large countries like Canada and the United States might have avoided potentially gargantuan national gazetteers by limiting them to names likely to be included on 1:250,000 maps.

15. Payne, "Development and Implementation."

16. Orth and Mangan, *Geographic Names and the Federal Government*, 35–36. Their titles misrepresented the first four lists as geographical dictionaries, a flaw Gannett quickly remedied by adopting the correct term, gazetteer, for the other eleven. An additional gazetteer is the 446-page *Geographic Dictionary of Alaska*, compiled by Marcus Baker and published in 1902. Baker's substantial discussion of the territory's geography and historical toponymy give him a stronger claim than Gannett to the title *Geographic Dictionary*.

17. Gannett, *Gazetteer of Colorado*, 122, 160.

18. Orth, "U.S. Board on Geographic Names." Gannett had been the Department of the Interior's representative on the board since 1890.

19. The end of the series coincided with a substantial expansion of the board's responsibilities in 1906 as well as other USGS deployments of Gannett, who served as geographer and assistant director for the Philippine census of 1903 and the Cuban census of 1907, and as geographer and editor for the National Conservation Commission between 1903 and 1909. For concise biographies of Gannett, see Darton, "Memoir of Henry Gannett"; and "Henry Gannett," *National Geographic*.

20. For a serials librarian's assessment, see Morehead, "Decision Lists." Currently the lists are published only electronically.

21. U.S. Geographic Board, *Sixth Report*, 20–21.

22. According to an annotated Library of Congress bibliography compiled

largely by Don Orth, the federal board's executive secretary for the Domestic Names Committee, "subsequent reorganization of the BGN in 1934 terminated the program." The state names committee in Massachusetts had completed a similar compilation, which "was never published." See Orth and Mangan, *Geographic Names and the Federal Government*, 8.

23. Heck, "Problem of a National Gazetteer."

24. Ibid., 234.

25. For a description of the role of fieldwork in harvesting toponyms for coastal charts of the U.S. Coast and Geodetic Survey, see Wraight, "Field Work in the U.S.C.&G.S."

26. Ibid., 238.

27. Heck and others, *Delaware Place Names*, 1.

28. Ibid., 79.

29. Ibid., 1. I estimated the total number of entries as 3,036 by extending the average for 11 pages, sampled systematically, to all 118 pages with names entries.

30. The 1918 map is the USGS edition of a World War I–era topographic map "surveyed in cooperation with the War Department." Pressing military needs account for lettering less elegant than usual for pre–World War II USGS topographic maps. The 1918 and 1955 maps had contour intervals of 10 feet, whereas the 1943 map had a contour interval of 20 feet, which might have been too coarse to pick up the low-lying island.

31. U.S. Geological Survey, *National Gazetteer of the United States of America— Delaware, 1983*, ix.

32. U.S. Geological Survey, *National Gazetteer of the United States of America— New Jersey, 1983*, ix.

33. U.S. Board on Geographical Names, *Decisions Rendered Between July 1, 1938 and June 30, 1939*, 29.

34. Ibid., 29. On p. 25, the *Decision List* includes a pair of variant cross-references for Nigger Pond: one to Potake Pond, New York, and the other to Cranberry Pond, New York.

35. An 1893 edition of the 1:62,500 map added a further complication: the feature labeled Nigger Pond on the 1910 map is missing. While it is possible that the pond was either man-made or formed after 1893 by a landslide or radical rise in the water table, a more likely explanation is sloppy fieldwork for the earlier map.

36. Roger Payne mentions these "intensive investigations" fleetingly in "Development and Implementation," 310–11. More recent experience suggests that the 40 percent estimate is high—for some areas the existing cartographic record might account for only a fifth of all feature names.

37. U.S. Geological Survey, *National Gazetteer of the United States of America, Concise*, ix.

38. Ibid. Orth's introduction to the Florida compilation says nothing to suggest that the series was ending.

39. See Payne, "Development and Implementation," esp. 310–13.

40. Abate, *Omni Gazetteer of the United States*.

41. The USGS names database is a basic resource for the Getty Thesaurus of

Geographic Names Online (http://www.getty.edu/research/conducting_research/vocabularies/tgn/), which offers worldwide as well as domestic coverage.

42. For further distinctions, see Payne, "Geographic Names Information System."

43. A Phase IA, often included in Phase I, harvested additional names from maps or charts published by the U.S. Forest Service, the National Park Service, and the National Ocean Service.

44. Roger Payne, writing in 1995, reported a projection completion date of 2005. See Payne, "Development and Implementation," 311.

45. Although most contracts cover a single state, the six New England states, shown as a block in figure 2.3, were treated as a single project. The number of bidders is small, and in recent years successful bidders typically had worked on previous state-level names compilation projects. Robin D. Worcester (systems manager, U.S. Geological Survey, Office of Geographic Names), in telephone discussion with the author, June 24, 2003.

46. Orth and Payne, *National Geographic Names Data Base,* 6.

47. Ibid., 22.

48. Ibid., 10–11.

49. Ibid., 6–7.

50. Julia Pastore (staff researcher, U.S. Geological Survey, Office of Geographic Names) and Robin D. Worcester, in telephone discussions with the author, June 26, 2003.

51. "Niger Post Office (historical)" is not listed in Foscue, *Place Names in Alabama.* The gazetteer's 175 pages include physical features. According to Robin Worcester, of the USGS Names Office, the contractor found the feature on a U.S. Post Office Department microform list of early nineteenth-century post offices. Robin D. Worcester, in telephone discussion with the author, June 26, 2003.

52. The creek is not listed in Miller, *Ohio Place Names.* This gazetteer's 286 pages focus almost exclusively on villages, towns, and other populated places. According to Robin Worcester, the contractor found the feature in a list developed by H. F. Raup, a geography professor at Kent State University. Robin D. Worcester, in telephone discussion with the author, June 26, 2003. Also see Raup, "The Names of Ohio's Streams."

53. Even so, "Niger Hill" is not to be found Espenshade, *Pennsylvania Place Names.*

54. Robin D. Worcester, e-mail communication with author, June 26, 2003.

55. Although the historical geography of toponyms is beyond the scope of this book, I refer interested readers to McArthur, "The GNIS and the PC"; Vasiliev, "Mapping Names"; and Zelinsky, "Along the Frontiers of Name Geography."

56. A remarkable scarcity of African Americans in western mining camps might be a factor; see Dickison, "Onomastic Amelioration," 16. However offensive to twentieth-century Americans, many *nigger* toponyms seem little more than an attempt to commemorate an African American who lived or died near the named feature. For example, Idaho's Nigger Brown Hill is "named for a black man who mined here and lost his life in a mining accident," and Nigger Creek commemo-

rates Jerry Greene, a teamster who "froze both his feet and died as a result." See Boone, *Idaho Place Names*, 271.

CHAPTER THREE

1. I also found eleven names containing *Chinkapin*, hardly a slur on persons of Chinese ancestry, as well as three equally irrelevant toponyms with *Chink* embedded in a longer word. One of the three official names I counted is Chinkhollow Creek, a California name likely to be a concatenation of Chink Hollow. For the three probably pejorative variants, *China* or *Chinese* had replaced *Chink* or *Chinks*.

2. I discussed the Negro Marsh and Wappingers Falls controversy in *Drawing the Line*, 54–57. Wappingers Falls was on a list of forty-five allegedly offensive names that the Department of Environmental Conservation submitted to the New York Committee on Geographic Names.

3. *American Heritage Dictionary*, 3rd ed., s.v. "wop."

4. See Conliff, "Wappinger"; and Turco, *Walks and Rambles*, 50–52.

5. Urbanek, *Wyoming Place Names*.

6. *American Heritage Dictionary*, 3rd ed., s.v. "dago."

7. Ibid., s.v. "kraut."

8. Most communities with a German name resisted the impulse to rename; see Rennick, "On the Success of Efforts."

9. I have seen other forms, but *Polack* is the preferred spelling; see *American Heritage Dictionary*, 3rd ed., s.v. "Polack."

10. Ibid., s.v. "gringo."

11. Espenshade, *Pennsylvania Place Names*. Gringo is a named collection of seven structures on the U.S. Geological Survey, Beaver, Pennsylvania, 15-minute topographic map, 1:62,500 (1904).

12. Dictionaries of slang and guidebooks to verbal insults largely ignore toponyms. Among other examples, I found no treatment of place names in Berrey and Van den Bark, *American Thesaurus of Slang*; Hughes, *Swearing*; and Jay, *Cursing in America*.

13. I found thirty-two official names containing *spade* as a separate word, and one variant, a Spade Branch that had been changed to Spadra Branch, perhaps to correct a misspelling.

14. After disallowing Ospook Creek, a stream in Alaska, I counted twenty-nine official names containing *spook* as a separate word, plus one variant, a Spook Lake that had been changed to Loon Lake.

15. I counted 827 official names that began with *coon* as a separate word and another 140 in which *coon*, also as a separate word, was preceded with one or more words like *little* or *north fork*. A strict focus on names with *coon* as a separate word ignored names with *raccoon* as well as names like Coontree Creek. Removing 60 obvious duplications for *coon* features in Logan County, Oklahoma, reduced the 140 count to 80 for a total of 1,027 official names. Variants raise the count to 1,082. I also found 67 first-place variants and another 13 in which one or more words precedes *coon*. Removing the 25 variants for which the official name also contains *coon* reduced the number of variants from 80 to 55.

16. Thorne, *Dictionary of Contemporary Slang*, 213–14.

17. Romig, *Michigan Place Names;* and Vogel, *Indian Names in Michigan.*

18. U.S. Library of Congress, *A World of Names*, 19.

19. Six *Mick* toponyms are official names and three are variants.

20. All of the seven *Scotchtowns* are official names. I counted 157 additional official names with *Scotch*, either separate or embedded, and 34 variants, only 14 of which refer to a feature with an official name not derived from *Scotch* or *Scot.*

21. See "A Talk with Jesse Jackson," *Newsweek;* and Schanberg, "Jackson as Polarizer."

22. Davis, "St. Simons Island."

23. "Ugh! Oops," *New Republic.*

24. *American Heritage Dictionary*, 3rd ed., s.v. "Jew."

25. I counted sixteen official toponyms and two variants, but disallowed "Jew Town," an obsolete spelling and redundant listing for the Georgia hamlet in question. GNIS also lists six features with names based on *Jews* (plural); Jews Quarter Island, now Dews Island (in North Carolina), is the only variant.

26. GNIS lists nine official names based on *jewfish* and one variant, Jewfish Key, now Fiesta Key.

27. Lynch, "Fishy Nickname." The goliath grouper is a threatened species; see National Oceanic and Atmospheric Administration, "Goliath grouper (*Epinephelus itajara*)."

28. *Chicago Tribune*, "Renaming the Jewfish"; and Roig-Franzia, "Coast to Coast."

29. Dahlberg, "Fish Lends Its Name"; and Sargeant, "Name Change Not Exactly a 'Goliath' Issue."

30. GNIS lists thirty features with official names containing *Chinaman* or *Chinamans*, and eight features with the word in a variant name. Among the variants, three reflect a change to merely another toponym incorporating *Chinaman* or *Chinamans*. I found no official or variant names based on *Chinamen.*

31. Urbanek, *Wyoming Place Names*, 38. For a selection of slightly different anecdotal explanations of the name—four of them, actually—see Whittlesey, *Yellowstone Place Names*, 35.

32. Lagerfeld, "Name That Dune"; and Willing, "'Cripple Creek.'"

33. Washington Post, "Negro Mountain Keeps Name"; and Papenfuse and others, *Maryland*, 60. My sources disagree over whether the battle occurred in 1758 or 1774, and whether Nemesis was fighting with Thomas Cresap or his son Michael, both of whom were actively involved in pushing the European frontier westward.

34. *Washington Post*, "Negro Mountain Keeps Name."

35. Stewart L. Udall to Edward P. Cliff, July 27, 1962, U.S. Board on Geographic Names, "Negro" file.

36. Edward P. Cliff to Stewart L. Udall, filed with minutes of October 16, 1962, meeting of the executive committee of the U.S. Board on Geographic Names. Quotation is from a carbon copy held by the U.S. Geological Survey, Office of Geographic Names; the copy might be the final version or a preliminary draft of the

board's reply to Udall. Julia Pastore (staff researcher, U.S. Board on Geographic Names), e-mail communication with author, June 13, 2003.

37. I counted thirty-one features with a name containing *Jap* as a whole word. None of these names are current. I reduced that count by eliminating Jap Hill (now Laderan Tangke) in the Northern Mariana Islands, now part of the United States as a commonwealth (like Puerto Rico) and thus within the purview of the Board on Geographic Names. GNIS also lists Japs Lake as a variant for Lake of the Woods, a large Minnesota feature. (Minnesota once had a *Jap* Lake, a smaller feature that is now Japanese Lake.) Of the remaining thirty *Jap* toponyms, all variants, only three (one each in Alabama, Florida, and West Virginia) lie in the east. Jap Post Office, in West Virginia, is a historical name listed without coordinates.

38. Minutes of the U.S. Board on Geographic Names, meeting of January 8, 1974.

39. For insights on the decision to imprison Japanese Americans but not German Americans or Italian Americans, see Robinson, *By Order of the President,* 4.

40. Because GNIS reports the year of change only for names individually endorsed by the federal board, I could not separate features affected by blanket renaming from those renamed before 1962. Of the 194 features with names that once contained *nigger, niger,* or *nigar,* only nine were changed by the board before 1962 and none were changed during 1962.

41. Although individual board action was unnecessary, if the board confirmed a name, the action was noted. Roger Payne (executive secretary, U.S. Board on Geographic Names), e-mail communication with author, July 21, 2003.

42. In contrast to the other three ethnic groups, African Americans have been prodigious in naming churches *African*—a specific name that accounts for 936 of the 990 toponyms containing *Africa* or *African.*

43. GNIS lists the embassies of these countries, but for consistency my counts do not include buildings.

44. Caldwell, "Renaming of Town."

45. Ibid.

46. I was deliberately conservative in omitting Spanish instances of *Negro,* which are common in the Southwest. Hence figure 3.4 probably overstates the number of toponyms in Arizona, California, and New Mexico that were intended to refer to African Americans.

47. These seven names are listed in the original version of House Bill 483, filed on March 12, but were replaced in the final version, passed unanimously by both houses in early June, and signed into law on June 23, 2003. North Carolina Legislature, Information/history, House Bill 483. A similar initiative emerged a year later in Florida; see Reeder, "Bill Targets Racially Sensitive Place Names."

48. Powell, *North Carolina Gazetteer,* 351. Also see *Greensboro (NC) News and Record,* "Bill Would Rid Maps of Offensive Names."

49. I first learned of the omissions from a *New York Times* reporter, who had already called the Geological Survey. The five toponyms were missing from the snapshot of GNIS I received in mid-May 2003. I updated my maps from the GNIS Web

site, which had the missing toponyms in early June. Geographic Names Office staff confirmed the additions, which are mentioned in the minutes of the June 12, 2003, meeting of the Domestic Names Committee. Although listed in a published gazetteer, at least three of the features were ill-defined; as of June 24, the GNIS records for Negro Skull Mountain, Negro Spring, and Negroskull Creek reported the latitude, longitude, and quadrangle name as "unknown"

50. For the historical roots of regional differences in boundary descriptions, see Thrower, *Original Survey and Land Subdivision*.

51. Belkin, "On Geographic Names."

52. Streeter, "Pressing for Change."

53. Franklin, "Cass Lake Students Want 'Squaw' Removed."

54. Minnesota State Senate, S.F. No. 574, 2nd Engrossment. The governor signed the bill into law on April 18, 1995.

55. Thus the law allows Squapan, which means "bears' den" in the local indigenous language.

56. Maine Legislature, "An Act Concerning Offensive Names."

57. Belluck, "A County Has a Word for It."

58. For the text of the law, see South Dakota Legislature, "Geographic Place Names Replaced." The bill was signed into law on March 2, 2001. Governor William Janklow was a strong supporter; see Associated Press, "Janklow's Efforts to Remove Offensive Words"; and Melmer, "Offensive Names."

59. Oregon Legislature, Senate Bill 488. Also see Pyati, "Legislators Vote to Cut 'Squaw.'"

60. Oregon Legislature, Senate Joint Memorial 3.

61. Idaho Legislature, House Concurrent Resolution no. 53.

62. The Women of Color Alliance gave the issue salience through public protests, including a March 2001 sit-in at the governor's office, as well as a documentary video, *Historical Impact of the S Word: From One Generation to the Next*, in which native women recount painful memories associated with *squaw*. See Taule, "Protestors of 'Squaw' Get No Answers"; and Rosario, "WOCA Builds Power."

63. Taule, "'The Ugliest Word I've Ever Heard.'"

64. Hall, "Actually, Hornbeck Did Defend the Word 'Squaw.'"

65. Associated Press, "Board Erases Slur."

66. "bull----" is probably *bullshit;* see Sahlberg, "What's in a Name?"

67. Associated Press, "Council Chairman Resigns."

68. For a concise survey of state-level efforts to ban *squaw* toponyms, see Gasque, "Structure and Controversy." Gasque mentions Montana as another state that passed a geographic names resolution.

69. See Schmitt, "Lake County Journal." Although local opposition seems to have delayed renaming for a few years, GNIS indicates that the federal board approved new names for Lake County's two *squaw* features in 2000.

70. Perhaps the classic example is the "water buffalo" incident at the University of Pennsylvania; see Kors and Silverglate, *The Shadow University*, 9–33.

71. Few assertions that *squaw* names carried no derogatory intent can be backed

up as effectively as an explanation offered for the name of Squaw Island in Orion Lake, near Orion, Michigan. According to a short, stereotype-laden article in *Historical Collections,* published by the Michigan Pioneer and Historical Society, the island was named for a group of Indian women who saved the lives of a white mill worker by confiscating the weapons of some young Indian males while they consumed a jug of whiskey. See Carpenter, "Squaw Island." For an example of *squaw* and the C-word, thinly disguised as "c***," see Adams, "Is Squaw an Obscene Insult?" Adams attributes the recent controversy over squaw names to a 1992 *Oprah Winfrey Show* on which Suzan Harjo noted the word's origin as a term for vagina.

72. As linguistics expert William Bright notes, "The derogatory use of 'squaw' has a background in racism, and racism is a continuing fact in American society." See Bright, "Sociolinguistics of the 'S-Word,'" esp. 214.

73. News accounts of anti-*squaw* efforts occasionally mention *papoose* place names as a secondary irritant; for an example, see Ferguson, "Drive to Erase Demeaning Names Rolls On."

74. *American Heritage Dictionary,* 3rd ed., s.v. "papoose."

75. See, for example, ibid., s.v. "pickaninny."

76. Elliott Moffat, quoted in the minutes of the State-Federal Roundtable, at the annual meeting of the Council of Geographic Names Authorities, Boise, Idaho, September 6, 2001.

77. Principle VI, approved by the Domestic Names Committee and the full board, was sent to the Department of the Interior for approval in January 2004. It allows only two options: change the name or declare it "historical." Roger Payne, telephone conversation with author, December 1, 2003.

CHAPTER FOUR

1. From the title of Dingman, "Naughty Names."

2. Writers' Program, *Pennsylvania,* 543.

3. Espenshade, *Pennsylvania Place Names,* 297.

4. Ibid., 296–97.

5. Baker and Carmony, *Indiana Place Names,* 57.

6. See Dingman, "Naughty Names," 3.

7. In compiling figure 4.1, I omitted Pass Tit Francois, a Louisiana bayou, because *tit* is a contraction for *petit* and the name merely commemorates someone called Little Francis.

8. For a colorful essay on Maine's colorful (or off-color) feature names, see Rutherford, "Censorship and Some Maine Place Names."

9. A GNIS history note reports that the name appeared on advance copies of the 1962 map but was removed from the final version. I was unable to locate a 1962 edition but found a 1952 version of the map with the feature name missing. According to a USGS researcher who revisited the case, the name was published on 1948, 1962, and 1997 editions of the map, but not on 1952 and 1958 editions. Eve Edwards (staff researcher, U.S. Board on Geographic Names), e-mail communication with author, October 16, 2003.

10. Although GNIS lists no other current names based on *whore,* there are nine

variants, including Bunkers Whore Ledge (now East Bunker Ledge, on the Maine coast) and Whore Creek (now the Lewes and Rehoboth Canal, in Delaware). The most intriguing replacement is South Carolina's Escaped Whore Swamp, now known as Scape Ore Swamp. If you want to soften a potentially offensive name, try a Cockney accent.

11. McArthur, *Oregon Geographic Names,* 790.

12. *Washington (D.C.) Star-News,* "People in the News," found in the U.S. Geological Survey, Office of Geographic Names case file for Whorehouse Meadow Naughty Girl Meadow, Oregon. Other writers were fascinated by the name; for examples, see Hook, *All Those Wonderful Names,* 220; Large, "Naughty Girl Meadow by Some Other Name"; and Schmitt, "Ultimate Arbiter of Hill and Vale."

13. Lewis L. McArthur to U.S. Geological Survey, Menlo Park, CA, December 18, 1971, U.S. Board on Geographic Names, case file for Whorehouse Meadow/Naughty Girl Meadow, Oregon. The file also contains a January 6, 1972, letter from Pacific Region engineer Roy F. Thurston to the chief topographic engineer, back east. "[Naughty Girl Meadow] was the only version found during our field investigation," Thurston maintained. "We were not aware of the variant until receipt of Mr. McArthur's letter."

14. Minutes of the 284th Meeting of the Domestic Names Committee, U.S. Board on Geographic Names, August 8, 1972.

15. Oregon Geographic Names Board Proposal, submitted December 3, 1982, by Lewis L. McArthur and approved by the state board on September 15, 1983. U.S. Board on Geographic Names, case file for Whorehouse Meadow/Naughty Girl Meadow, Oregon.

16. U.S. Board on Geographic Names, case brief (domestic) for Whorehouse Meadow, Oregon, approved December 8, 1983.

17. Oxford English Dictionary Online, http://dictionary.oed.com s.v. "dildo."

18. For a history of the development and evolution of the electromechanical vibrator, see Maines, *The Technology of Orgasm,* quotation on 121. Another instance of a word with a wholly different meaning in contemporary English occurs in the Austrian village of Fucking, apparently named to commemorate a sixth-century chap named Focko. Suspicious of a hoax, I verified its existence by consulting the National Geospatial-Intelligence Agency geographic names database (http://gnswww.nga.mil/geonames/GNS/index.jsp).

19. "A Dilemma in Dildo," *Maclean's;* and Rayburn, "Unfortunate Connotations Acquired by Some Canadian Toponyms." *Maclean's* reported the number of signatures as "about 400 of the fishing village's 950 residents," whereas Rayburn put the count at "about 350 of the 800 area residents." The small number of houses in the village proper (fig. 4.5) suggests that the enterprising Elford was casting a wide net in his effort to have the name changed. Like Dildodians, residents of the Austrian village of Fucking rejected a proposal to change the name. See *Sunday Tribune* (Ireland), "Fucking Villagers Vote against Name Change." Although annoyed by frequent thefts of the road sign identifying their community, residents value the traditional name and its attraction to Anglophone tourists. Also see Lerch, "Many VT-er's Most Desired Travel Destination."

20. The Web page could be a spoof, but it lists an inn, a funeral home, and a Lions Club that appears to operate as a bar or café. See St. Georges High School, Dildo, Newfoundland, K–12 Internet On Ramp, Newfoundland Educational Networking Group, "Dildo Days."

21. Rayburn, *Naming Canada*, 80.

22. Governor Mario M. Cuomo, Executive Order no. 113.

23. Marc C. Gerstman, deputy commissioner and general counsel, DEC, to Edmund Winslow, chair of the New York State Committee on Geographic Place Names, August 25, 1989. Philip Lord, who succeeded Winslow, provided correspondence related to this issue.

24. Green, *Newspeak*, 69–70.

25. New York State Committee on Geographic Place Names, Policies and Procedures (draft).

26. *Kofa* is an acronym for a branch of the famous King Ranch—King *of* Arizona; see Dingman, "Naughty Names," 2.

27. *Webster's Third New International Dictionary*, unabridged, s.v. "shite."

28. Ibid., s.v. "shitten."

29. Ibid., s.v. "shitepoke."

30. Rayburn, *Naming Canada*, 82.

31. Ibid., 83.

32. GNIS lists a second variant, Isla Vista County Park, which suggests an administrative feature the name of which is controlled by a local agency, not the federal board.

33. GNIS uncovered forty-one streams officially named Cripple Creek as well as eight other *cripple* names like Cripple Bush Creek and Cripple Cow Creek; *cripple* toponyms can be found in twenty-one states. The likelihood of negative reaction is apparent in objections to a 1996 proposal to christen an unnamed stream in Texas Cripple Creek. Curiously, a local judge who used a wheelchair and wanted to honor people with similar disabilities had requested the change, which the state names board rejected after receiving objections from groups representing the disabled. See Siegel, "Name Changes for Landmarks."

CHAPTER FIVE

1. San Miguel, "How Is Devils Tower a Sacred Site?"

2. Gunderson, *Devils Tower*, 31–32. Gunderson describes other legends, some quite different, that distinguish Devils Tower as a holy place.

3. Tim Giago, quoted in Brooke, "What's in a Name?"

4. San Miguel, "How Is Devils Tower a Sacred Site?"

5. Smith, "Tribes Say Devils Tower Is No Name for a Pious Peak."

6. See, for example, Associated Press, "Cubin Resurrects Bill to Keep Devils Tower Name"; and Associated Press, "Enzi, Thomas Seek to Keep Devils Tower Name."

7. Quotation from Policy I: Names Being Considered by Congress; see Orth and Payne, *Principles, Policies, and Procedures*, print version, as revised in 2003. This policy avoids a potential conflict insofar as acts of Congress and executive orders

take precedence over the board's decisions. The phrase "and the Executive Branch" was added to Policy I in 2003 to correct an oversight—the reference had previously been used in Principle III. Roger Payne (executive secretary, U.S. Board on Geographic Names), e-mail communication with author, December 8, 2003.

8. The highest mountain in North America, Denali is part of the Alaska Range. According to linguistics expert James Kari, the mountain has two names, which reflect different writing systems. Athabaskans living north of the range call the mountain "the high one," while those living south of the range refer to it as "big mountain." Kari, "The Tenada-Denali–Mount McKinley Controversy."

9. H.R. 164, 108th Cong., 1st sess., http://thomas.loc.gov.

10. Quoted in Loewen, *Lies across America*, 53.

11. Dickey, "Discoveries in Alaska," quoted in Moore, *Mt. McKinley*, 14–15.

12. Stewart, *American Place-Names*, 270.

13. Moore, *Mt. McKinley*, 17, 179. Moore's source is Browne, *The Conquest of Mount McKinley*, 8.

14. Moore, *Mt. McKinley*, 1.

15. For a facsimile of the Wrangel map, see ibid., 3.

16. Quoted in Moore, *Mt. McKinley*, 9. I could not find the source of this quotation—Moore was somewhat lax in his citations—but it is consistent with other observations by Brooks. For examples, see Brooks, "An Exploration to Mt. McKinley," esp. 441–43; Brooks, *The Mount McKinley Region, Alaska*, 25–27; and Brooks, *Blazing Alaska's Trails*, 291–93.

17. Moore, *Mt. McKinley*, 8–9; and National Park Service, "Denali National Park and Preserve Mountaineering History."

18. Dickey, "The Sushitna River, Alaska."

19. Muldrow, "Mount McKinley."

20. Stuck, *The Ascent of Denali*, vii. Stuck was not alone in his ascent; his expedition included Walter Harper, Harry Karstens, and Robert Tatum.

21. National Park Service, "Denali National Park and Preserve Mountaineering Booklet."

22. According to annotation to title 16, chapter 1, subchapter XXXIX, sec. 351 of the *United States Code*, "'Denali National Park' [was] substituted in text for 'Mount McKinley National Park' pursuant to Pub. L. 96–487, Sec. 202(3)(a), which is classified to section 410hh-1(3)(a) of this title and which added lands to the park, established additional land as the Denali National Preserve, and redesignated the whole as the Denali National Park and Preserve." Pub. L. 96–487, title II, sec. 202(3)(a), December 2, 1980, 94 Stat. 2382; online at http://www4.law.cornell.edu/uscode/16/351.notes.html. I was unable to determine who initiated the change, a four-line insertion buried in the technical language of the 181-page Alaska National Interest Lands Conservation Act. Because title II addressed the National Park System, it seems likely that some National Park Service personnel were aware of the change.

23. Orth, "The Mountain Was Wronged"; and *New York Times*, "What's in a Name?"

24. The left-hand map in figure 5.2 is an excerpt from the online road map

offered by the Trade and Development Division at http://www.dced.state.ak.us/ trade/tou/learn/roadmap.htm. The right-hand map in figure 5.2 is an excerpt from the "Interior" map, one of five regional maps at http://www.dced.state.ak.us/ trade/tou/learn/region.htm.

25. Jo Antonson (state historian, Alaska), e-mail communication with author, November 12, 2003.

26. See Merriam and McCormick, *Report of the United States Geographic Board on S. J. Res. 64;* Orth, "The Mountain Was Wronged," 431–32; Reese, *Pierce County Name Origins,* 123–44; and Stewart, *Names on the Land,* 364–72. The Reese title is a mimeographed document in the collection of the U.S. Geological Survey Library, Reston, Virginia.

27. Merriam and McCormick, *Report of the United States Geographic Board on S. J. Res. 64,* 1.

28. Ibid., 4, 7–8, 8.

29. Orth, "The Mountain Was Wronged," 432.

30. For the text of Policy X: Names of Native American Origin, approved in 1996, see Orth and Payne, *Principles, Policies, and Procedures* (1997).

31. Ibid., Policy X, sec. 1. This phrase survived an extensive rewriting of the Native American names policy in 2003 and is now embedded in Policy X, sec. 3.

32. Quotation is from the pre-2003 version of Principle I: Use of the Roman Alphabet, which has been revised to read: "Official domestic geographic names are written in Roman script. Diacritical and special marks, however, may be added to names as specified in Policy VI: Use of Diacritical Marks." Orth and Payne, *Principles, Policies, and Procedures* (2003).

33. The strong recommendation for a generic element is stated in ibid., Policy X, sec. 7; identical wording is now a part of Policy X, sec. 6. Orth and Payne, *Principles, Policies, and Procedures* (2003).

34. In the board's pre-2003 policy document, the issue of sovereignty on tribal land was addressed by Policy X, sec. 3 and 4, in ibid. The relevant policy, rewritten as Policy X, sec. 1, para. 1, is reinforced by Executive Order no. 13175, signed by President William J. Clinton on November 6, 2000. See Clinton, "Consultation and Coordination with Indian Tribal Governments."

35. Roger Payne, e-mail communication with author, December 4, 2003.

36. Policy X, sec. 1, para. 2, as rewritten in 2003, in Orth and Payne, *Principles, Policies, and Procedures.*

37. Ibid.

38. Ibid., Policy X, sec. 2, as rewritten in 2003.

39. U.S. Board on Geographic Names, *Decision List 1999,* 20.

40. Orth and Payne, *Principles, Policies, and Procedures* (1997), Policy X, sec. 6. The quotation survives in Policy X, sec. 5(b), as rewritten in 2003, in Orth and Payne, *Principles, Policies, and Procedures.* Both the old and the new policy documents define "linguistically appropriate" by requiring that "the language from which the name is derived is, or once was, spoken in the area by a relatively permanent population."

41. State names authorities report difficulty in eliciting a timely response from

tribal councils asked to comment on proposed names. Officials in several states recommend calling rather than writing. Minutes of the State-Federal Roundtable, at the annual meeting of the Council of Geographic Names Authorities, Salt Lake City, Utah, September 6, 1996.

42. Roger Payne, in telephone conversation with author, December 1, 2003. Payne has been executive secretary of the full board since 1993.

43. Orth and Payne, *Principles, Policies, and Procedures,* Policy X, sec. 1, as rewritten in 2003.

44. Not all native names in the northeastern states are pleasant sounding and easy to pronounce; see Whitbeck, "Geographic Names in the United States."

45. GNIS lists Chargoggagoggmanchauggagoggchaubunagungamaugg, meaning, "You fish on your side, I'll fish on my side," as one of twenty-six variants for a feature officially known as Lake Chaubunagungamaug.

46. U.S. Board on Geographic Names, *Decision List 1997,* 10.

47. Cutler, "The Battle the Indian Won."

48. U.S. Board on Geographic Names, *Decision List 1998,* 8.

49. U.S. Board on Geographic Names, *Decision List 1999,* 29.

50. Ibid., 24.

51. Dalyrymple, "What's in a Name?" Ki-a-Cut Falls, the spelling supported by the environmental group that sponsored the original petition, is listed in GNIS as a variant.

52. Sullivan, "Honoring, and Unearthing, Indian Place Names."

53. Basso, *Wisdom Sits in Places,* 43.

54. U.S. Board on Geographic Names, *Field Investigation of Native American Place Names.*

55. Kapono, "Hawaiian Language Renaissance."

56. For insights on the simultaneous suppression and appropriation of native culture, see Herman, "The Aloha State"; and Wood, *Displacing Natives,* esp. 10–13.

57. Roger Payne, in telephone conversation with author, December 1, 2003. Fallacious fusion is especially troubling to linguists because at least half of all Hawaiian place names consist of multiple words even before appending an English generic like *stream,* often redundant to native speakers. See Pukui, Elbert, and Mookini, *Place Names of Hawaii,* 243–44.

58. Kapono, "Hawaiian Language Renaissance."

59. The statement on official languages survives intact in the revised state constitution that took effect on January 1, 2000. Article XV, sec. 4, states, "English and Hawaiian shall be the official languages of Hawaii, except that Hawaiian shall be required for public acts and transactions only as provided by law." Constitution of the State of Hawaii, http://www.hawaii.gov/lrb/con/conart15.html.

60. Quotation is from Policy VII: Use of Diacritical Marks, approved in 1986; see Orth, *Principles, Policies, and Procedures* (1989), 17.

61. Officials at Haleakala National Park requested the change in early 1994, according to Roger Payne, in telephone conversation with author, December 1, 2003.

62. Quotation is from Policy VI: Use of Diacritical Marks, in Orth and Payne,

Principles, Policies, and Procedures (2003). Easing of restrictions on diacritical marks also removed a key obstacle to including Hopi toponyms; see Rundstrom, "The Role of Ethics, Mapping, and the Meaning of Place."

63. Orth and Payne, *Principles, Policies, and Procedures* (1997, 2003), Policy VI.

64. U.S. Board on Geographic Names, *Decision List 1999*.

65. For the official sanctioning of the name, see U.S. Board on Geographic Names, *Decision List no. 8902*, 2. In 1999, at the request of the Hawaii names board, Puʻu ʻŌʻo. became the official name. Jennifer Runyon (staff researcher, U.S. Board on Geographic Names), e-mail communication with author, December 5, 2003. According to GNIS, the feature is in Hawaii Volcanoes National Park, ten miles east of Kīlauea Crater. For information on the eruption and an example of the name's Hawaiian spelling, see U.S. Geological Survey, Hawaiian Volcano Observatory, "Summary of the Puʻu ʻŌʻo."

66. Annual decision lists were not published after 1999 because of staffing and budget cuts. In 2000, 2001, 2002, and 2003, the board approved 752, 679, 740, and 380 Hawaiian names, respectively. Roger Payne, in telephone conversation with author, December 1, 2003. For a less optimistic view of USGS policy and its victimization of the Hawaiian people, see Louis, "The Authoritative Textualization of Hawaiian Place Names." Writing in 1998, Renee Louis condemns the "inherent arrogance" of applied toponomy. Ibid., 57. Also see Louis, "Indigenous Hawaiian Cartographer."

67. To expedite map revision, the Hawaii board forwarded lists of new names simultaneously to the federal board in Washington and the Geological Survey's mapping center in Denver. Roger Payne, in telephone conversation with author, December 1, 2003.

68. Ibid.

69. Pukui, Elbert, and Mookini, *Place Names of Hawaii*, 203.

70. Ibid., 250. Ulehawa Beach Park, on the 1998 map, illustrates one advantage of not translating native names into English, insofar as only a native speaker would know that Ulehawa means "filthy penis." Ibid., 214–15.

71. Roger Payne, in telephone conversation with author, December 1, 2003.

72. Published in 1974, the second edition of *Place Names of Hawaii* remains the most authoritative published source of native toponyms. See Pukui, Elbert, and Mookini, *Place Names of Hawaii*. In 1989 the press published an abridged version by the same authors. Both are still in print.

73. This URL directs browsers to a Web site hosted on GeoCities.com. For the pronunciation guide, see Hawaiian Language Home Page. "Hawaiian Language Pronunciation: A List of Common Mispronunciations."

74. U.S. Geographic Board, *Sixth Report*, 72–73. Pronunciation came up again in the mid-1980s, when the board appointed a committee that studied the issue for two years, noted the need to select from among several competing systems for describing pronunciation, and recommended that the federal board not become involved. In addition to ruling out any attempt to standardize pronunciation, the board officially withdrew the 1932 pronunciation guide in 1985. Roger Payne, in conversation with the author, July 29, 2003.

75. See the British Columbia Ministry of Sustainable Resource Management, "Nisga'a Names, Nisga'a Lands."

76. Carlson, McHalsie, and Perrier, *A Stó:lō Coast Salish Historical Atlas.* The Stó:lō Nation, which holds the copyright, is listed as a copublisher and sponsor.

77. Ibid., 2.

78. Facts about the new territory are from the Government of Nunavut Web site, http://www.gov.nu.ca/.

79. Müller-Wille, "Inuit Toponymy and Cultural Sovereignty"; and Rayburn, "Native Names for Native Places." For additional insights on Inuit toponymy and cartography, see Rundstrom, "A Cultural Interpretation of Inuit Map Accuracy."

80. Rayburn, *Naming Canada,* 136–40.

81. The small-scale Nunavut map on the Atlas of Canada Web site has a 2001 copyright date; see Natural Resources Canada, Atlas of Canada, "Nunavut, 2001."

82. Although French became the official language for public business in 1974, with the passage of Bill 22, coverage was broadened substantially in 1977 by Bill 101, which included advertising, business names, product inserts, and public signage. For insights, see Coulombe, "Making Sense of Law 101."

83. The Geographical Names Board of Canada has official jurisdiction over eighty major features with broad, pan-Canadian significance and well-established names like Hudson Bay, Lake Huron, and Vancouver Island (and their French equivalents Baie d'Hudson, Lac Huron, and Île de Vancouver). For further information on the Canadian names board, see Kerfoot and Rayburn, "The Roots and Development of the Canadian Permanent Committee on Geographical Names"; and Kerfoot, "United States and Canada." For the board's guidelines and procedures, see Natural Resources Canada, Canadian Geographical Names.

84. René, "Current Use Determines Name."

85. Geographer Anna Nieminen explored the controversy's many facets in her doctoral dissertation; see Nieminen, "The Cultural Politics of Place Naming in Québec."

86. Peritz, "Quebec Names Islands to Mark Bill 101."

87. Roslin and Webb, "Iles 101."

88. Bearskin quoted in ibid.

CHAPTER SIX

1. Kazuhiko Koshikawa (director, Japan Information Center, New York), letter to author, May 29, 2003.

2. See, for example, *Toronto Star,* "Korea Out to Rename the Sea of Japan"; and WuDunn, "Koreans, Still Bitter, Recall 1945."

3. The list was published in 1928 as International Hydrographic Bureau, *Limits of Oceans and Seas.* Second and third editions were published in 1937 and 1953, respectively. According to Rear Admiral Kenneth Barbor, director of the IHB, draft fourth editions circulated for comment in 1986 and 2002 were not published because a majority of member states would not certify their approval. Kenneth Barbor, e-mail communication with author, January 8, 2004. Also see *Toronto Star,* "No Safe Middle Ground in the Map Wars"; and Reid, "S. Koreans Almost Sink Sea of

Japan Plan." News reports typically refer to the International Hydrographic Organization, which is now the official name of the consultative organization formed in 1921 as the International Hydrographic Bureau. The organization grew from nineteen member states in 1921 to seventy-three member states in 2003. Its headquarters is in Monaco, at the International Hydrographic Bureau. As the IHO Web site (http://www.iho.shom.fr/) explains, the IHB is the secretariat of the IHO.

4. Ibison, "Japan Protests about Changing of Sea's Name"; and *Daily Yomiuri* (Tokyo), "Japan, ROK Dispute Sea Name."

5. See *Daily Yomiuri* (Tokyo), "Japan, ROK Dispute Sea Name."

6. Ministry of Foreign Affairs of Japan, "Sea of Japan," 6. Also see Ledyard, "Cartography in Korea," esp. 249.

7. Koshikawa, letter to author.

8. Lee, "East Sea in World Maps."

9. Ibid. In identifying cartographer John Senex as "Jone [sic] Senex," Lee's list raises questions about its compiler's reliability.

10. Ministry of Foreign Affairs of Japan, "Sea of Japan."

11. Study results are reported in Kim, "'East Sea' on a Roll."

12. Lee, "East Sea in World Maps"; and *Korea Times*, "US University Library Lists Old Maps."

13. Results are attributed to a report from the South Korean news agency Seoul Yonhap, summarized in the Financial Times Global News Wire, "Korean, Chinese Scholars Say East Sea Was Predominant Term in the West."

14. For Korean accounts of their successes, see Lee, "East Sea in World Maps"; and Korean Overseas Information Service, "Rediscover the Proper Name for Korea's East Sea."

15. I confirmed the use of East Sea by two online references in early 2004. The *Oxford Atlas* adopted East Sea between its ninth and tenth editions, published in 2001 and 2002, respectively. Rand McNally began using both names simultaneously in some of its products after receiving a letter from the public affairs officer at the South Korean Embassy in March 1996; see Bowers, "Embassy Row." The firm's *Premier World Atlas* added East Sea parenthetically for the 1997 printing, but the twentieth edition of *Goode's World Atlas,* published in 2000, ignores the toponym. As with several of its many atlases, Rand McNally does not number editions of its *Premier World Atlas,* a title that the Research Libraries Group bibliographic database suggests was dormant between 1981 and 1997.

16. *National Geographic Atlas of the World,* 7th ed., plate 104. The patch was offered as a JPEG file with a short accompanying explanation as National Geographic Society, "Sea of Japan (East Sea)." A higher resolution, 300 dots-per-inch PDF version was also provided. I was unable to confirm Korean reports of an earlier change, in 1999; for other National Geographic Society maps, see Korean Overseas Information Service, "Rediscover the Proper Name for Korea's East Sea." The society launched the seventh edition of its world atlas in October 1999, and the following month a *Christian Science Monitor* interview with several of its cartographic staff indicated that the East Sea was still an open question; see Chaddock, "A Good Map Is (Easy) to Find."

17. National Geographic Society, "Sea of Japan (East Sea)."

18. National Geographic Society, "Keep Your Atlas Up to Date."

19. Rosenberg, "Sea of Japan vs. East Sea." I found Rosenberg's revised map at http://geography.about.com/library/cia/blcsouthkorea.htm.

20. For the CIA's unadulterated take on the Sea of Japan, see U.S. Central Intelligence Agency, "Korea, South."

21. Schmitt, "North Korea MIG's Intercept U.S. Jet."

22. Han, letter to the editor. For another example of a complaint, see *Toronto Star*, "No Safe Middle Ground in the Map Wars."

23. Keleny, "Mea Culpa."

24. Jones, "Whose Sea Is It?"

25. Demick, "A 'C' Change in Spelling Sought for the Koreans"; and Shaffer, "Korea, Corea, Gorea, or Koria?"

26. George F. Cram Company, *Cram's Unrivaled Atlas of the World*, 173. The Sea of Japan appears on the following page, on a separate map of Japan.

27. Not everyone uses Beijing, but Peking is rare these days, except perhaps in restaurants, where "Beijing duck" might seem a bit strange. China officially changed Peking to Beijing on January 1, 1979, when it adopted a new transliteration system. My "just a decade" estimate is based on the *Christian Science Monitor*, which officially switched over ten years later; see Tyson, "Why Peking Becomes Beijing."

28. The list also included 174 variants; see U.S. Board on Geographic Names, *First Report of the United States Board on Geographic Names*, 43–46.

29. U.S. Board on Geographic Names, *First Report on Foreign Geographic Names*. The estimate of twenty-five hundred names is from Burrill, "Reorganization," 649.

30. U.S. Board on Geographic Names, *First Report on Foreign Geographic Names*, 55, 74, 89.

31. Ibid., 56, 71.

32. For insights on the history and role of the Foreign Names Committee, see Burrill, "Reorganization"; Randall, "The U.S. Board on Geographic Names and International Programs"; and Randall, "The U.S. Board on Geographic Names and Its Work in Foreign Areas."

33. The GEOnet Names Server provides access to downloadable issues of the *Foreign Names Information Bulletin*. See National Geospatial-Intelligence Agency, GEOnet Names Server. Although the *Bulletin*'s cover page indicates that publication is "on a quarterly basis, or as needed," the government typically publishes fewer than four issues per year.

34. For a concise history of the gazetteers and a sample of their contents, see Randall, *Place Names*, 115–17, 179–80.

35. For an overview of federal policies on foreign names, see Dillon, "Recent Discussion"; and Randall, *Place Names*, 97–113.

36. For additional information on translation, transliteration, transcription, and romanization, see Randall, *Place Names*, esp. 109–12; and Aurousseau, *Rendering of Geographical Names*, esp. 50–60, 99–112.

37. Aurousseau, *Rendering of Geographical Names*, 56–57.

38. For further information, see Permanent Committee on Geographical Names for Official British Use.

39. Aurousseau, *Rendering of Geographical Names*, 106.

40. Randall, *Place Names*, 77–78.

41. U.S. Board on Geographic Names, *First Report on Foreign Geographic Names*, 18.

42. Randall, "The U.S. Board on Geographic Names and International Programs," 252.

43. Randall, *Place Names*, 110.

44. For information on UN standardization efforts, see ibid., 81–89; Randall, "What's in a (Place) Name?"; and United Nations Statistics Division, "Geographical Names: Overview," http://unstats.un.org/unsd/geoinfo/about_us.htm.

45. For a list of the nine working groups, see United Nations Statistics Division, "Geographical Names: UNGEGN Working Groups." For the definition of exonym, see United Nations Group of Experts on Geographical Names, *Glossary*, 10.

46. Resolution VIII/5 of the 2002 conference, titled "Working Group on Exonyms of the United Nations Group of Experts on Geographical Names," sets the working group's terms of reference; see United Nations Group of Experts on Geographical Names, "Working Group on Exonyms," 36–37. For a concise summary of earlier UN interest in exonyms, see Breu, "Exonyms."

47. Sievers and Jordan, "International Workshop on Exonyms." School atlases might have contributed to this increased use of exonyms; see Sandford, "The Geographical Name in Modern School Atlases."

48. Between 1967 and 1998, the UN names-standardization conference adopted 167 resolutions; for the full text of these resolutions see Geographical Names Board of Canada, "Resolutions Adopted."

49. Sievers and Jordan, "International Workshop on Exonyms."

50. For insights on "donor" and "receiver" principles, see Geelan, "Standardization of Geographical Names."

51. When the National Geographic Society revised its world atlas in the early 1990s, one-third of the maps required changes; see Barringer, "Chaos Spells Trouble for Map Makers."

52. See Kamm, "Conflict in the Balkans."

53. Mark Athanasios C. Karras to senior editors, *National Geographic*, December 9, 1992. Copy of letter provided by a source within the National Geographic Society.

54. U.S. Department of State, "Guidance Bulletin No. 12."

55. Thomas, "US Policy on Name of Burma."

56. U.S. Department of State, "Background Note: Burma."

57. *New York Times*, "Burma Out, Myanmar In."

58. A particularly pointed design uses diagonal stripes identified in the map key as representing "Israeli Occupied Territory"; see, for example, *World Atlas of Nations*, 86. A typical solution when colors represent elevation categories is a prominent dashed border surrounding the label West Bank, as in *Philip's Concise World Atlas*, 75.

59. Dates in this paragraph were verified using Cohen, *Columbia Gazetteer of the World*. Also see Nainan, "'Mumbai' to Replace Bombay on Map."

60. For insights on toponymic politics in the former Soviet Union, see Murray, *Politics and Place-Names*.

61. Safire, "Bring Back Upper Volta."

62. Allen, "Byelorussia, Hello Belarus."

63. Miller, "It's All in a Name."

CHAPTER SEVEN

1. U.S. Central Intelligence Agency, "The Disputed Area of Kashmir."

2. McCarthy, "Geography"; and Microsoft Corporation, "Microsoft's Geopolitical Product Strategy Team."

3. Encyclopaedia Britannica Online, s.v. "Amman," http://www.search.eb.com/eb/article?tocId=9007193 (accessed January 6, 2005).

4. The Population Reference Bureau Web site put the island's mid-2003 population at 900,000; see Population Reference Bureau, DataFinder. Although Northern Cyprus has not had a census since 1978, an official Web site reported the 2000 population as 210,047, which might be a bit high as well as questionably precise. See Turkish Republic of Northern Cyprus, "Introductory Survey."

5. Reunification of Cyprus is a goal of the European Union, which admitted it in hope of rapprochement. See Peel, "Trouble Is Brewing for Europe over Cyprus." Turkey, itself eager for EU membership, has encouraged Turkish Cypriots to cooperate. See Sachs, "With Eye on Europe."

6. Hoge, "Cyprus Greeks and Turks Agree on Plan."

7. For a fuller historical background to the Cyprus conflict, see Minority Rights Group, *Cyprus;* and Solsten, *Cyprus.*

8. Tartter, "National Security," esp. 219.

9. Estimated numbers of refugees vary between 170,000 and 200,000 Greek Cypriots and between 40,000 and 60,000 Turkish Cypriots. For examples, see Meleagrou and Yesilada, "The Society and Its Environment," esp. 80, 89; and Calotychos, "Interdisciplinary Perspectives," esp. 8.

10. Tartter, "National Security," 221.

11. See, for example, Barchard, "Turkey Unperturbed by Hellenic Uproar."

12. United Nations Sub-Commission on the Prevention of Discrimination, "Violations of Human Rights in Cyprus." For similar sentiments, see United Nations General Assembly, "Resolution 37/253, adopted May 16, 1983."

13. Tassos Papadopoulos (president of Republic of Cyprus), speech at official dinner, November 25, 2003.

14. Republic of Cyprus Press and Information Office, "Results of Invasion."

15. Ibid.

16. "Cyprus," *Economist.*

17. O'Dwyer, "Candle in the Sea."

18. Hamilos, "Comment and Analysis."

19. Huggler, "No Children to Hear the Village Bell."

20. Also see Kadmon, *Toponymy*, 80–81.

21. Ladbury and King, "Settlement Renaming in Turkish Cyprus."

22. Ibid., 364.

23. Culcasi, "Cartographic Representations of Kurdistan in the Print Media," 43; and Izady, *The Kurds*, 61.

24. Kadmon, *Toponymy*, 80.

25. Ibid., 82.

26. Ibid.

27. Benvenisti, *Sacred Landscape*, 54.

28. Katz, "Identity, Nationalism, and Placenames," esp. 105–6.

29. Government of Palestine, *Transliteration . . . of Personal and Geographical Names for Use in Palestine*.

30. Letter from Izhak Ben-Zvi to the general secretary of the government of Palestine, quoted in Katz, "Identity, Nationalism, and Placenames," 108.

31. Cohen and Kliot, "Israel's Place-Names as Reflections of Continuity and Change."

32. Ibid.

33. Cohen and Kliot, "Place-Names in Israel's Ideological Struggle."

34. Benvenisti, *Sacred Landscape*, 45.

35. Kadmon, *Eretz Israel*, quoted in ibid., 45.

36. Benvenisti, *Sacred Landscape*, 45.

37. Ibid., 20.

38. Ibid., 20–21; also see Bar-Am, "In the Middle of Nowhere." Questions also arose about the correct location of Mount Sinai; see Fishkoff, "Looking for Mount Sinai."

39. Benvenisti cites Central Zionist Archive KKL/5/20503 as the source for this quote but provides no further details; see Benvenisti, *Sacred Landscape*, 32.

40. Hourani and Heyda, *Gazetteer of Israel*, 150.

41. For insights on the site's uncertainty, see MacDonald, *East of the Jordan*, 45–61.

42. For a concise cartographic summary of the evolution of Israel and its border, see U.S. Central Intelligence Agency, *The Gaza Strip and West Bank*.

43. Maps and tables listing village names and reasons for abandonment are in Morris, *Birth of the Palestinian Refugee Problem*, viii–xx; quotation on xiv.

44. Ibid., 237.

45. For a concise description of topographic maps for Israel, see Parry and Perkins, *World Mapping Today*, 314–16.

46. Ibid., xiv–xv.

47. Benvenisti, *Sacred Landscape*, 43.

48. See, for example, Deir Yassin Remembered; and the MidEast Web Maps, Palestine maps. Also see MidEast Web Maps, "The Israeli Camp David II Proposals for Final Settlement."

49. Deir Yassin Remembered. Revisionist Israeli historian Benny Morris confirms the date and the slaughter: "The IZL/LHI attack on Deir Yassin near Jerusalem on 9 April ended not only in a massacre but also in the expulsion by the conquering unit of the surviving Arab villagers." See Morris, *1948 and After*, 84.

50. I slightly inflated an estimate, reported in 1998 dollars, as "nearly $24 billion." See Kershner, "The Refugee Price Tag." This figure does not include the lesser but not inconsiderable value of property left behind by Jewish immigrants from Arab states. See Rischbach, "An Answer for Jews and Palestinians." Neither group of refugees has been compensated.

51. Fischbach, *Records of Dispossession*, esp. 338–39.

CHAPTER EIGHT

1. For an analysis of the widespread cultural significance of naming streets after Martin Luther King Jr., especially in the Southeast, see Alderman, "A Street Fit for a King."

2. *Globe and Mail*, "There's No Place Like Home."

3. County planning director Karen Kitney and senior assistant corporation counsel Thomas Carnrike quoted in English, "(Doing the) James Street Shuffle."

4. Beer, "Council Tables Plan"; Associated Press, "Renaming Highway after Tubman Stopped"; and Shaw, "Tubman Dedication Set for May 27."

5. The mayor, who supported the change, is one of the council's five voting members. MapQuest.com shows a "Tubman Ln" but the DeLorme Mapping's *Upstate New York City Street Maps* does not.

6. Quindlen, "A Day for Renaming Places."

7. *New York Times*, "Politics Intrude"; and *New York Times*, "Corrections."

8. Breasted, "Street Name Change Stirs a Tempest."

9. *New York Times*, "14 Arrested in Protest"; and Anderson and Dunlap, "Mayor in a Hurry."

10. Zelinsky, "The Game of the Name," 44. These trends are neither new nor obsolete; see Minton, "Names of Real-Estate Developments"; and Norris, "Unreal Estate."

11. Zelinsky, "The Game of the Name," 45.

12. Rodriguez, "Soul of a New Neighborhood," 74.

13. Shuger, "Los Angeles Postcard."

14. Ibid., 7.

15. According to the introductory paragraph on GNIS in the domestic names manual, "All names in the database, except for variant names, are considered official for Federal use, by either Board policy or decision or under the procedures of the organization responsible for its administrative names." See Orth and Payne, *Principles, Policies, and Procedures* (online edition).

16. *Locale* is defined in appendix C of the Feature Class Definitions in U.S. Geological Survey, *Geographic Names Information System Data Users Guide 6*.

17. U.S. Geological Survey, "Frequently Asked Questions about GNIS."

18. Because I considered only locales with *mall* or *shopping* as part of their name, my counts underreport retail centers, especially in Clark County, where names with the generic *plaza* are common. In the American West in particular, a plaza is often a cluster of stores or offices.

19. I discuss trap streets, and Gould Street in particular, in Monmonier, "Map

Traps." For the court decisions, see *Feist Publications, Inc. v. Rural Telephone Service Co.*; and *Alexandria Drafting Co. v. Andrew H. Amsterdam d/b/a Franklin Maps.*

20. For an overview of sovereignty claims, see Joyner and Theis, *Eagle Over the Ice*, 36–40, 225.

21. Stokke, "The Making of Norwegian Antarctic Policy," esp. 386.

22. Australian Antarctic Division, Classroom Antarctica, "International Sovereignty, Unit 7.1."

23. Ibid.

24. For discussion of the history and significance of the treaty, see Stokke and Vidas, *Governing the Antarctic.*

25. U.S. Central Intelligence Agency, "Antarctic Region."

26. International law experts Christopher Joyner and Ethel Theis identified the period from the mid-1930s to the International Geophysical Year, "when the United States encouraged its nationals to claim territory on its behalf," as the third stage in America's evolving attitude toward Antarctica. See Joyner and Theis, *Eagle Over the Ice*, 37.

27. For examples, see Miller, "Planetabling from the Air"; Joerg, "The Topographical Results of Ellsworth's Trans-Antarctic Flight of 1935"; and "New Maps of the Antarctic." For a concise overview of mapping in Antarctica, see Parry and Perkins, *World Mapping Today*, 562–67.

28. U.S. Board on Geographical Names, *The Geographical Names of Antarctica.* A revised edition published in 1956 was titled *Geographic Names of Antarctica: Official Standard Names Approved by the U.S. Board on Geographic Names.*

29. U.S. Office of Geography, *Antarctica.*

30. Alberts, *Geographic Names of the Antarctic.*

31. Alberts, *Geographic Names of the Antarctic*, 2nd ed.

32. As of mid-May 2004, GNIS had 13,660 official Antarctic names and 4,418 variants. Roger Payne (executive secretary, U.S. Board on Geographic Names), e-mail communication with author, May 10, 2004.

33. Aurousseau, "The Treatment of Antarctic Names," 487. Aurousseau was secretary of the Permanent Committee on Geographical Names for British Official Use.

34. Ibid., 490.

35. U.S. Geological Survey, "Policy Covering Antarctic Names," http://geonames .usgs.gov/antex.html. For an earlier but largely similar statement of the policy, see Alberts, *Geographic Names of the Antarctic*, vii–xix.

36. U.S. Geological Survey, "Policy Covering Antarctic Names."

37. Ibid.

38. U.S. Geological Survey, GNIS Antarctic Query Web page.

39. See Cervellati and others, "A Composite Gazetteer of Antarctica"; and Programma Nazionale di Ricerche in Antartide, Composite Gazetteer of Antarctica."

40. SCAR's role is purely advisory insofar as it does not adjudicate controversies or set international policy.

41. Recommendation XXV-7 of the Working Group on Geodesy and Geo-

graphic Information, adopted in 1998 by delegates to SCAR's twenty-fifth annual meeting, is noted in Cervellati and others, "A Composite Gazetteer of Antarctica."

42. Names decisions within the twelve-mile limit of the territorial sea are within the purview of the Domestic Names Committee, which consults the appropriate state-level names authority for features within three miles of the coastline. Manuscript for revised draft (undated) furnished by Roger Payne around December 2003, of Orth and Payne, *Principles, Policies, and Procedures,* 35.

43. GNS provides little guidance to retrievals for undersea features, identified only the feature code U (for undersea); see National Geospatial-Intelligence Agency, GEOnet Names Server.

44. For an account of Tharp and Heezen's discovery, see Lawrence, *Upheaval from the Abyss.* Also see "New Undersea Features"; "Ocean Floor Discoveries in the Atlantic"; and "Uncharted Undersea Mountains."

45. Randall, *Place Names,* 92.

46. Sources differ on several of the dates reported in this paragraph; see Jones, "Historical Background to GEBCO"; National Oceanic and Atmospheric Administration, "History of GEBCO, 1903–2003"; Monahan, "The General Bathymetric Chart"; and Newson, "The General Bathymetric Chart."

47. Newson, "The General Bathymetric Chart," 42. According to Newson, the first edition provided only "a few major ocean-bed and terrestrial place names."

48. Monahan, "The General Bathymetric Chart," 15.

49. General Bathymetric Map of the Oceans (GEBCO), "Names of Undersea Features." The gazetteer can be downloaded as either a PDF file or an Excel spreadsheet.

50. "General Principles Governing the Naming of New Small Oceanic Features."

51. "Nomenclature of Ocean Bottom Features."

52. Intergovernmental Oceanographic Commission and International Hydrographic Organization, *Standardization of Undersea Feature Names,* 2–16 to 2–30.

53. Alfred Wegener Institute for Polar and Marine Research, "Fram Strait Feature Names."

54. Intergovernmental Oceanographic Commission and International Hydrographic Organization, *Standardization of Undersea Feature Names,* 2–2. According to GEBCO's "Names of Undersea Features," Molloy helped explore the north polar seas in the 1950s, 1960s, and 1970s. I could not determine whether he was still alive in 2003, but in 1960 Molloy and four colleagues executed the first submerged transit of the Northwest Passage by way of Parry Channel. See U.S. Navy, Arctic Submarine Laboratory, "Historical Timeline."

55. Intergovernmental Oceanographic Commission and International Hydrographic Organization, *Fifteenth Meeting,* 19. ACUF's secretary suggests that these comments, not identified as a consensus of the SCUFN membership, might be merely the sentiments of a single committee member that whoever prepared the annual report inserted without attribution. Trent C. Palmer (secretary, Advisory Committee on Undersea Features), e-mail communication with author, May 17, 2004.

56. Intergovernmental Oceanographic Commission and International Hydrographic Organization, *Fifteenth Meeting*, 19.

57. U.S. Navy, Naval Meteorological and Oceanographic Command, "Landry Bernard Retires as NAVO Technical Director."

58. Intergovernmental Oceanographic Commission and International Hydrographic Organization, *Fifteenth Meeting*, 19.

59. Ibid.

60. Intergovernmental Oceanographic Commission and International Hydrographic Organization, *Standardization of Undersea Feature Names*, guideline II.A.5 on 2–2; and U.S. Board on Geographic Names, *Policies and Guidelines for the Standardization of Undersea Feature Names*, guideline II.A.5 on 2.

61. Intergovernmental Oceanographic Commission and International Hydrographic Organization, *Sixteenth Meeting*, 27. On the same page, the 2003 report echoes an earlier complaint by objecting to "the frequent to routine award of specific names for significant open ocean seafloor topographic entities for retiring U.S. agency, military service or commercial personalities, or for currently newsworthy individuals."

62. Intergovernmental Oceanographic Commission and International Hydrographic Organization, *Fifteenth Meeting*, 21.

63. Randall, *Place Names*, 93.

64. Trent C. Palmer, e-mail communication with author, May 17, 2004.

65. Intergovernmental Oceanographic Commission and International Hydrographic Organization, *Sixteenth Meeting*, 18.

66. Although Sagan used "billions and billions" as a book title, he claimed never to have spoken the phrase, made popular by parodists like Johnny Carson. See Sagan, *Billions and Billions*, 13–15.

67. For a concise history of the IAU, see Trimble, "What, and Why, Is the International Astronomical Union?" Also see Kross, "What's in a Name?"

68. Strobell and Masursky, "Planetary Nomenclature," esp. 98. For information on Gilbert's chart, see Whitaker, *Mapping and Naming the Moon*, 10–15. For a list of milestones in planetary mapping, see Batson, Whitaker, and Williams, "History of Planetary Cartography," table on 55.

69. Batson, Whitaker, and Williams, "History of Planetary Cartography," 12–13. Batson and his coauthors call van Langren's chart the "first useful map" of the Moon (13).

70. Proctor, *Other Worlds than Ours*, 107–10.

71. Batson, Whitaker, and Williams, "History of Planetary Cartography," 34–35.

72. Strobell and Masursky, "Planetary Nomenclature," 103–5.

73. For a summary of relevant cartographic techniques, see Batson, "Cartography."

74. Examples include Andersson and Whitaker, *NASA Catalogue of Lunar Nomenclature;* Batson, Bridges, and Inge, *Atlas of Mars;* and Greeley and Batson, *The NASA Atlas of the Solar System.*

75. Based on an equatorial radius of 3,396 km, I estimate the area cover by the excerpt in figure 8.6 as 189,000 km^2 or 73,000 mi^2.

76. See the table "Categories for Naming Features on Planets and Satellites" in Strobell and Masursky, "Planetary Nomenclature," 100–101; and U.S. Geological Survey, Astrogeology Research Program, Gazetteer of Planetary Nomenclature.

77. For further information on the naming of asteroids, also called minor planets, see Burdick, "Name That Star!"

78. U.S. Geological Survey, Astrogeology Research Program, "Eros Nomenclature."

79. Strobell and Masursky, "Planetary Nomenclature," 102; and Whitaker, *Mapping and Naming the Moon*, 178–80.

80. Strobell and Masursky, "Planetary Nomenclature," 137–38; and U.S. Geological Survey, Astrogeology Research Program, "IAU Rules and Conventions."

81. U.S. Geological Survey, Astrogeology Research Program, "IAU Rules and Conventions," Rule 6.

82. The Board on Geographic Names set up an Advisory Committee on Extraterrestrial Features in 1975, but the committee has been dormant since the mid-1980s. Randall, *Place Names*, 95.

83. Counts are from U.S. Geological Survey, Astrogeology Research Program, "IAU Working Group and Task Group Members." In May 2004 Americans chaired three of the six task groups.

84. According to the agency that services the WGPSN, "If the members of the task group agree that the name is appropriate, it can be retained for use when there is a request from a member of the scientific community that a specific feature be named." U.S. Geological Survey, Astrogeology Research Program, "How Names Are Approved."

85. International Star Registry, "Name a Star."

86. The Copyright Office registers a copyright for the book, not each star name therein. Registration is a straightforward process insofar as the Copyright Office will register a copyright for any publication for which the applicant sends along two copies, asserts authorship and originality, fills out the form properly, and pays the $30 registration fee.

87. A "Virtual Star package" includes a certificate that can be printed online; see Name a Star, Inc., "Virtual Star Package."

88. International Astronomical Union, "Naming Stars."

EPILOGUE

1. Harley, "Cartography, Ethics and Social Theory," 4.

2. Nicolaisen, "Placenames and Politics."

3. Ormeling, "Minority Toponyms."

4. For examples, see Nijhuis, "Tribal Immersion Schools Rescue Language and Culture"; and Pierre, "Northwest Tribe Struggles to Revive Its Language."

5. For examples, see Briggs and Carter, "Tribes Racing to Save Dying Languages"; Fishman, *Reversing Language Shift;* Hornberger, *Indigenous Literacies in the Americas;* and Schwartz, "Speaking Out and Saving Sounds to Keep Native Tongues Alive."

Bibliography

Abate, Frank R., ed. *Omni Gazetteer of the United States of America*. 11 vols. Detroit: Omnigraphics, 1991.

About Geography. "Korea, South." http://geography.about.com/library/cia/blcsouthkorea.htm.

Adams, Cecil. "Is Squaw an Obscene Insult?" *Chicago Reader*, March 17, 2000. http://www.straightdope.com/columns/000317.html.

Alaska Department of Community and Economic Development. Regional maps. http://www.dced.state.ak.us/trade/tou/learn/region.htm.

Alberts, Fred G., ed. *Geographic Names of the Antarctic: Names Approved by the United States Board on Geographic Names*. Washington, DC: National Science Foundation, 1981.

———. *Geographic Names of the Antarctic: Names Approved by the United States Board on Geographic Names*. 2nd ed. Washington, DC: National Science Foundation, 1995.

Alderman, Derek H. "A Street Fit for a King: Naming Places and Commemoration in the American South." *Professional Geographer* 52 (2000): 672–84.

Alexandria Drafting Co. v. Andrew H. Amsterdam d/b/a Franklin Maps, 43 U.S.P.Q.2d 1247 (June 4, 1997).

Alfred Wegener Institute for Polar and Marine Research. "Fram Strait Feature Names." http://www.awi-bremerhaven.de/GEO/Bathymetry/framstr/framfeat.html.

Allen, Henry. "Byelorussia, Hello Belarus: Lost among the Geo-Linguistic Labels of the New World Order." *Washington Post*, January 1, 1992.

Allord, Gregory J., and Kerie J. Hitt. "Linking Digital Technology to Printing Technology for Producing Publication-Quality Color Graphics." In *Selected Papers in the Applied Computer Sciences, 1990*. U.S. Geological Survey Bulletin 1908, B1–B7. Washington, DC: U.S. Government Printing Office, 1990.

Anderson, Susan Heller, and David W. Dunlap. "Mayor in a Hurry." *New York Times,* December 22, 1984.

Andersson, Leif E., and Ewen A. Whitaker. *NASA Catalogue of Lunar Nomenclature*. Washington, DC: National Aeronautics and Space Administration, Scientific and Technical Information Branch, 1982.

Associated Press. "Board Erases Slur from Federal Maps," September 6, 2001.

———. "Council Chairman Resigns in Protest of Political Correctness," August 7, 2002.

———. "Cubin Resurrects Bill to Keep Devils Tower Name," February 6, 1999.

———. "Enzi, Thomas Seek to Keep Devils Tower Name," July 22, 1999.

———. "Janklow's Efforts to Remove Offensive Words from Place Names Stir Hope," April 10, 2000.

———. "Renaming Highway after Tubman Stopped on Technicality," December 5, 2003.

Aurousseau, Marcel. *The Rendering of Geographical Names*. London: Hutchinson University Library, 1957.

———. "The Treatment of Antarctic Names." *Geographical Review* 38 (1948): 487–90.

Australian Antarctic Division, Classroom Antarctica. "International Sovereignty, Unit 7.1." http://www.classroomantarctica.aad.gov.au/pdfs/International_U7_1_Sov_SO.pdf.

Baker, Marcus. *Geographic Dictionary of Alaska*. U.S. Geological Survey Bulletin 187. Washington, DC: U.S. Government Printing Office, 1902.

Baker, Ronald L., and Marvin Carmony. *Indiana Place Names*. Bloomington: Indiana University Press, 1974.

Bar-Am, Aviva. "In the Middle of Nowhere." *Jerusalem Post,* November 23, 2001.

Barchard, David. "Turkey Unperturbed by Hellenic Uproar." *Financial Times* (London), March 10, 1984.

Barringer, Felicity. "Chaos Spells Trouble for Map Makers." *Chicago Tribune,* December 29, 1991.

Basso, Keith H. *Wisdom Sits in Places: Landscape and Language among the Western Apache*. Albuquerque: University of New Mexico Press, 1996.

Batson, Raymond M. "Cartography." In *Planetary Mapping*, edited by Ronald Greeley and Raymond M. Batson, 60–95. Cambridge: Cambridge University Press, 1990.

Batson, Raymond M., P. M. Bridges, and Jay L. Inge. *Atlas of Mars: The 1:5,000,000 Map Series*. Washington, DC: National Aeronautics and Space Administration, Scientific and Technical Information Branch, 1979.

Batson, Raymond M., Ewen A. Whitaker, and Don E. Williams. "History of Planetary Cartography." In *Planetary Mapping*, edited by Ronald Greeley and Raymond M. Batson, 12–59. Cambridge: Cambridge University Press, 1990.

Beaman, W. M. "Topographic Mapping." In *Topographic Instructions of the United*

States *Geological Survey,* edited by C. H. Birdseye. U.S. Geological Survey Bulletin 788, 161–378. Washington, DC: U.S. Government Printing Office, 1928.

Beeler, Madison S. "George R. Stewart, Toponymist." *Names* 24 (1976): 77–85.

Beer, Beth. "Council Tables Plan to Rename Route 20." *Syracuse Post-Standard,* May 23, 2003.

Belkin, Lisa. "On Geographic Names and Cleaning Them Up." *New York Times,* February 14, 1990.

Belluck, Pam. "A County Has a Word for It; Problem Is, It's 'Moose.'" *New York Times,* February 21, 2002.

Benvenisti, Meron. *Sacred Landscape: The Buried History of the Holy Land Since 1948.* Translated by Maxine Kaufman-Lacusta. Berkeley: University of California Press, 2000.

Berrey, Lester V., and Melvin Van den Bark. *The American Thesaurus of Slang.* 2nd ed. New York: Thomas Y. Crowell, 1953.

Birdseye, C. H., ed. *Topographic Instructions of the United States Geological Survey.* U.S. Geological Survey Bulletin 788. Washington, DC: U.S. Government Printing Office, 1928.

Boone, Lalia. *Idaho Place Names: A Geographical Dictionary.* Moscow: University of Idaho Press, 1988.

Bower, Nancy. "How'd Those Yellowstone Places Get Their Names?" *Idaho Falls Post Register,* October 12, 1997.

Bowers, Paige. "Embassy Row: Sea Change." *Washington Times,* August 28, 1997.

Bowley, Diana. "'Moose' on the Loose: Feds Take Issue with Surge in Spots Names for Antlered Animal." *Bangor (ME) Daily News,* December 24, 2001.

Breasted, Mary. "Street Name Change Stirs a Tempest in Melting Pot." *New York Times,* May 4, 1976.

Breu, Josef. "Exonyms." *World Cartography* 18 (1986): 30–32.

Briggs, Kara, and Steven Carter. "Tribes Racing to Save Dying Languages." *Sunday Oregonian,* December 9, 2001.

Bright, William. "The Sociolinguistics of the 'S-Word': Squaw in American Placenames." *Names* 48 (2000): 207–16.

British Columbia Ministry of Sustainable Resource Management, Land Information BC. "Nisga'a Names, Nisga'a Lands." http://srmwww.gov.bc.ca/bcnames/g2_nl.htm.

Brooke, James. "What's in a Name? An Affront, Say Several Tribes." *New York Times,* November 17, 1996.

Brooks, Alfred Hulse. *Blazing Alaska's Trails.* 2nd ed. Fairbanks: University of Alaska Press, 1973.

———. "An Exploration to Mt. McKinley, America's Highest Mountain." *Journal of Geography* 2 (1903): 441–69.

———. *The Mount McKinley Region, Alaska.* U.S. Geological Survey Professional Paper 70. Washington, DC: U.S. Government Printing Office, 1911.

Browne, Belmore. *The Conquest of Mount McKinley.* New York: G. P. Putnam's Sons, 1913. Reprint, Cambridge, MA: Riverside Press, 1956.

Burdick, Alan. "Name That Star!" *Discover,* February 2000.

Burrill, Meredith F. "Reorganization of the United States Board on Geographic Names." *Geographical Review* 35 (1945): 647–52.

Caldwell, Earl. "Renaming of Town Divides Negroes on Coast." *New York Times,* December 26, 1968.

Calotychos, Vangelis. "Interdisciplinary Perspectives: Difference at the Heart of Cypriot Identity and Its Study." In *Cyprus and Its People: Nation, Identity, and Experience in an Unimaginable Community, 1955–1997,* edited by Vangelis Calotychos, 1–32. Boulder, CO: Westview Press, 1998.

Carlson, Keith Thor, Albert (Sonny) McHalsie, and Jan Perrier. *A Stó:lō Coast Salish Historical Atlas.* Vancouver, BC: Douglas and McIntyre, 2001.

Carpenter, C. K. "Squaw Island—How It Received Its Name." *Historical Collections* 13 (1889): 486–88.

Cervellati, Roberto, and others. "A Composite Gazetteer of Antarctica." *SCAR Bulletin,* no. 138 (July 2000). http://www.scar.org/Publications/bulletins/webbull_138.htm.

Chaddock, Gail Russell. "A Good Map Is (Easy) to Find." *Christian Science Monitor,* November 9, 1999.

Chicago Tribune. "Renaming the Jewfish," July 23, 2001, North Sports Final Edition.

Clinton, William J. (president). "Consultation and Coordination with Indian Tribal Governments." Executive Order no. 13175, November 6, 2000. *Federal Register* 65 (2000): 67249–52.

Cohen, Saul B., ed. *The Columbia Gazetteer of the World.* New York: Columbia University Press, 1998.

Cohen, Saul B., and Nurit Kliot. "Israel's Place-Names as Reflections of Continuity and Change in Nation-Building." *Names* 29 (1981): 227–48.

———. "Place-Names in Israel's Ideological Struggle over the Administered Territories." *Annals of the Association of American Geographers* 82 (1992): 653–80.

Conliff, Steven Edwin Konkapot. "Wappinger." In *Gale Encyclopedia of Native American Tribes,* edited by Sharon Malinowski and others, 1:323–26. Detroit: Gale, 1998.

Cook, Karen Severud. "The Historical Role of Photomechanical Techniques in Map Production." *Cartography and Geographic Information Science* 29 (2002): 137–53.

Coulombe, Pierre A. "Making Sense of Law 101 in the Age of the Charter." *Quebec Studies* 17 (Fall–Winter 1993): 45–58.

Council of Geographic Names Authorities. Minutes of the State-Federal Roundtable, annual meeting, September 6, 2001, Boise, Idaho.

Culcasi, Karen Leigh. "Cartographic Representations of Kurdistan in the Print Media." Master's thesis, Syracuse University, 2003.

Cuomo, Mario M. (governor). New York State Executive Order no. 113, signed October 31, 1988.

Cutler, Charles L., Jr. "The Battle the Indian Won." *American Illustrated History* 6 (January 1972): 20–27.

"Cyprus: A Bitter Lemon Squeezed Dry." *Economist,* September 4, 1976.

Dahlberg, John-Thor. "Fish Lends Its Name, Name Gives Offense." *Los Angeles Times*, December 1, 2002.

Daily Yomiuri (Tokyo). "Japan, ROK Dispute Sea Name," February 6, 2002.

Dalyrymple, Helen. "What's in a Name?" *Library of Congress Information Bulletin* 59 (February 2000): 38–41.

Darton, N. H. "Memoir of Henry Gannett." *Annals of the Association of American Geographers* 7 (1917): 68–70.

Davis, Jingle. "St. Simons Island: Black Owners Hold Tight to Land." *Atlanta Constitution*, December 20, 2001.

Deir Yassin Remembered. http://www.deiryassin.org/.

DeLorme Mapping Co. *Upstate New York City Street Maps*. Freeport, ME: DeLorme, 1990.

Demick, Barbara. "A 'C' Change in Spelling Sought for the Koreans." *Los Angeles Times*, September 13, 2003.

Dickey, William A. "Discoveries in Alaska." *New York Sun*, January 24, 1897.

———. "The Sushitna River, Alaska." *National Geographic* 8 (1897): 322–27.

Dickison, Roland. "Onomastic Amelioration in California Place Names." *Names* 16 (1968): 13–18.

"A Dilemma in Dildo." *Maclean's*, August 19, 1985.

Dillon, Leo. "Recent Discussion in the United States Board on Geographic Names Concerning the Creation of Anglicized Exonyms." Eighth United Nations Conference on the Standardization of Geographical Names, Berlin, 27 August–5 September 2002. Provisional agenda item 10, E/CONF.94/CRP.78. http://unstats.un.org/unsd/geoinfo/N0243895.pdf.

Dingman, Lester F. "Naughty Names: Geographic Names—Colorful or Offensive?" In *Naughty Names*, edited by Fred Tarpley, 1–4. Commerce, TX: South Central Names Institute, 1975.

Elsasser, Glen. "Name That Place: Where America's Geographic Features Get a Title and Court of Appeal." *Chicago Tribune*, January 21, 1991.

English, Molly. "(Doing the) James Street Shuffle." *Syracuse New Times*, January 10, 2001. http://newtimes.rway.com/2001/011001/shakin.shtml.

Espenshade, Abraham H. *Pennsylvania Place Names*. State College: Pennsylvania State College, 1925.

Evans, R. T., and others. *History of the Topographic Branch (Division): Predecessor Surveys*, preliminary draft. Washington, DC: U.S. Geological Survey, 1975.

Federal Writers' Project. *Florida: A Guide to the Southernmost State*. New York: Oxford University Press, 1939.

Feist Publications, Inc. v. Rural Telephone Service Co., 499 U.S. 340 (1991).

Ferguson, Dean. "Drive to Erase Demeaning Names Rolls On." *Lewiston (ID) Morning Tribune*, February 11, 2003.

Financial Times Global News Wire. "Korean, Chinese Scholars Say East Sea Was Predominant Term in the West," October 15, 2003.

Fischbach, Michael R. *Records of Dispossession: Palestinian Refugee Property and the Arab-Israeli Conflict*. New York: Columbia University Press, 2003.

Fishkoff, Sue. "Looking for Mount Sinai." *Jerusalem Post,* September 27, 1996.

Fishman, Joshua A. *Reversing Language Shift: Empirical and Theoretical Foundations of Assistance to Threatened Languages.* Clevedon, UK: Multilingual Matters, 1991.

Foscue, Virginia O. *Place Names in Alabama.* Tuscaloosa: University of Alabama Press, 1989.

Franklin, Robert. "Cass Lake Students Want 'Squaw' Removed from Names of Lakes, City." *Minneapolis Star Tribune,* June 4, 1994.

French Hydrographic Service (SHOM). The International Hydrographic Organization (IHO). http://www.iho.shom.fr/.

Gannett, Henry. *A Gazetteer of Colorado.* U.S. Geological Survey Bulletin 291. Washington, DC: U.S. Government Printing Office, 1906.

———. *A Manual of Topographic Operations.* U.S. Geological Survey Monograph XXII. Washington, DC: U.S. Government Printing Office, 1893.

Gasque, Thomas J. "Structure and Controversy: What Names Authorities Adjudicate." *Names* 48 (2000): 199–206.

Geelan, P. J. M. "Standardization of Geographical Names." *Geographical Journal* 138 (1972): 390–92.

Gehrke, Robert. "Piestewa Peak Likely to Wait Five Years for Federal Name Change." Associated Press, May 8, 2003.

General Bathymetric Map of the Oceans (GEBCO). "Names of Undersea Features." http://www.ngdc.noaa.gov/mgg/gebco/underseafeatures.html.

"General Principles Governing the Naming of New Small Oceanic Features." *International Hydrographic Review* 32 (November 1955): 157.

Geographical Names Board of Canada. "Resolutions Adopted at the Seven United Nations Conferences on the Standardization of Geographical Names." List of resolutions, 2001. http://unstats.un.org/unsd/geoinfo/unresolutions.htm.

George F. Cram Company. *Cram's Unrivaled Atlas of the World.* Grand New Census Edition. Rochester, NY: W. H. Stewart, 1891.

Giago, Tim. "Walk Up to Any Indian Woman and Call Her a Squaw." *San Diego Union-Tribune,* May 2, 2003.

Globe and Mail (Toronto). "There's No Place Like Home," May 25, 2003.

Goddard, Ives. "Since the Word Squaw Continues to Be of Interest." *News from Indian Country,* mid-April 1997.

Goldman, John J. "Arizona's Squaw Peak Is Renamed to Honor Soldier." *Los Angeles Times,* April 19, 2003.

Government of Nunavut. Official Web site. http://www.gov.nu.ca/.

Government of Palestine. *Transliteration from Arabic and Hebrew into English, from Arabic into Hebrew, and from Hebrew into Arabic with Transliterated Lists of Personal and Geographical Names for Use in Palestine.* Jerusalem, 1931.

Greeley, Ronald, and Raymond Batson. *The NASA Atlas of the Solar System.* Cambridge: Cambridge University Press, 1997.

Green, Jonathon. *Newspeak: A Dictionary of Jargon.* London: Routledge and Kegan Paul, 1989.

Greensboro (NC) News and Record. "Bill Would Rid Maps of Offensive Names," March 13, 2003.

Gunderson, Mary Alice. *Devils Tower: Stories in Stone.* Glendo, WY: High Plains Press, 1988.

Hall, Bill. "Actually, Hornbeck Did Defend the Word 'Squaw.'" *Lewiston (ID) Morning Tribune,* March 3, 2002.

Hamilos, Paul. "Comment and Analysis." *Guardian* (London), August 25, 2003.

Han, Eung Soo. Letter to the editor, *Chicago Sun-Times,* September 13, 1998.

Harley, J. B. "Cartography, Ethics and Social Theory." *Cartographica* 27 (Summer 1990): 1–23.

Haskel, Daniel, and John Calvin Smith. *A Complete Descriptive and Statistical Gazetteer of the United States.* New York: Sherman and Smith, 1843.

Hawaii Constitution, art. XV, sec. 4. http://www.hawaii.gov/lrb/con/conart15 .html.

Hawaiian Language Home Page. "Hawaiian Language Pronunciation: A List of Common Mispronunciations." http://www.geocities.com/~olelo/ wl-mispronunciations.html.

Heck, Lewis. "The Problem of a National Gazetteer." *Names* 1 (1953): 233–38.

Heck, L. W., and others. *Delaware Place Names.* U.S. Geological Survey Bulletin 1245. Washington, DC: U.S. Government Printing Office, 1966.

Heinen, Tom. "Squaw: A Word to Be Banned?" *Milwaukee Journal Sentinel,* December 22, 1996.

"Henry Gannett." *National Geographic* 26 (1914): 609–13.

Herman, R. D. K. "The Aloha State: Place Names and the Anti-Conquest of Hawai'i." *Annals of the Association of American Geographers* 89 (1999): 76–102.

Hoge, Warren. "Cyprus Greeks and Turks Agree on Plan to End 40-Year Conflicts." *New York Times,* February 14, 2004.

Hook, J. N. *All Those Wonderful Names: A Potpourri of People, Places, and Things.* New York: John Wiley and Sons, 1991.

Hornberger, Nancy H., ed. *Indigenous Literacies in the Americas: Language Planning from the Bottom Up.* Berlin: Mouton de Gruyter, 1997.

Hourani, May M., and Charles M. Heyda. *Gazetteer of Israel.* 2nd ed. Washington, DC: Defense Mapping Agency, 1983.

Huggler, Justin. "No Children to Hear the Village Bell." *Independent* (London), July 30, 1999.

Hughes, Geoffrey. *Swearing: A Social History of Foul Language, Oaths and Profanity in English.* Oxford: Blackwell, 1991.

Ibison, David. "Japan Protests about Changing of Sea's Name." *Financial Times* (London), August 16, 2002.

Idaho Legislature. House Concurrent Resolution no. 53, March 2002. http://www .leg.state.or.us/orlaws/sess0600.dir/0652ses.html

Institute of American Indian Studies. "Arizona Pressured to Outlaw 'Squaw.'" News bulletin, December 9, 1999. http://www.usd.edu/iais/bulletin/12bull.html.

Intergovernmental Oceanographic Commission and International Hydrographic Organization. *Fifteenth Meeting of the GEBCO Sub-Committee on Undersea Feature Names (SCUFN): Summary Report.* Monaco: International Hydrographic Bureau, 2002.

————. *Sixteenth Meeting of the GEBCO Sub-Committee on Undersea Feature Names (SCUFN): Summary Report.* Monaco: International Hydrographic Bureau, 2003.

————. *Standardization of Undersea Feature Names: Guidelines, Proposal Form, Terminology.* Bathymetric Publication no. 6, 3rd ed., English/Spanish version. Monaco: International Hydrographic Bureau, 2001.

International Astronomical Union. "Naming Stars." http://www.iau.org/IAU/FAQ/starnames.html.

International Hydrographic Bureau. *Limits of Oceans and Seas.* Special Publication no. 23. Monaco: International Hydrographic Bureau, 1928.

International Star Registry. "Name a Star." http://www.starregistry.com.

Izady, Mehrdad R. *The Kurds: A Concise Handbook.* Washington, DC: Crane Russak, 1992.

Jay, Timothy. *Cursing in America: A Psycholinguistic Study of Dirty Language in the Courts, in the Movies, in the Schoolyards and on the Streets.* Philadelphia: John Benjamins, 1992.

Joerg, W. L. G. "The Topographical Results of Ellsworth's Trans-Antarctic Flight of 1935." *Geographical Review* 26 (1936): 454–62.

Jones, David. "Whose Sea Is It?" *Washington Times,* December 1, 2002.

Jones, Meirion T. "Historical Background to GEBCO." Appendix to *User Guide to the Centenary Edition of the GEBCO Digital Atlas and Its Data Sets.* Liverpool, UK: Natural Environment Research Council, 2003. http://www2.ifjf.uib.no/gmtdata/pdf/manual.pdf

Joyner, Christopher C., and Ethel R. Theis. *Eagle Over the Ice: The U.S. in the Antarctic.* Hanover, NH: University Press of New England, 1997.

Julyan, Robert. "Protecting the Endangered Blank Spots on Maps: The Wilderness Names Policy of the United States Board on Geographic Names." *Names* 48 (2000): 217–26.

Kadmon, Naftali. *Eretz Israel,* no. 22. Jerusalem: Bialik Institute, 1992.

————. *Toponymy: The Lore, Laws and Language of Geographical Names.* New York: Vantage Press, 2000.

Kamm, Henry. "Conflict in the Balkans: Macedonia; For Greeks, It Is More than a Name." *New York Times,* April 23, 1994.

Kapono, Eric M. "Hawaiian Language Renaissance." In *Atlas of Hawai'i,* 3rd ed., edited by Sonia P. Juvik, James O. Juvik, and Thomas R. Paradise, 199. Honolulu: University of Hawai'i Press, 1998.

Kari, James. "The Tenada–Denali–Mount McKinley Controversy." *Names* 34 (1986): 347–51.

Katz, Yossi. "Identity, Nationalism, and Placenames: Zionist Efforts to Preserve the Original Local Hebrew Names in Official Publications of the Mandate Government of Palestine." *Names* 43 (1995): 103–18.

Keleny, Guy. "Mea Culpa: Why I Had to Get My Teeth into Robert Fisk." *Independent* (London), March 21, 2003.

Kennedy, Randall. *Nigger: The Strange Career of a Troublesome Word.* New York: Pantheon Books, 2002.

Kerfoot, Helen. "United States and Canada: Partners in Geographic Names Standardization." *Names* 48 (2000): 243–48.

Kerfoot, Helen, and Alan Rayburn. "The Roots and Development of the Canadian Permanent Committee on Geographical Names." *Names* 38 (1990): 183–92.

Kershner, Isabel. "The Refugee Price Tag." *Jerusalem Report,* July 17, 2000.

Kim, Min-hee. "'East Sea' on a Roll in Replacing 'Sea of Japan.'" *Korea Herald,* March 13, 2003.

Koeman, C. "The Application of Photography to Map Printing and the Transition to Offset Lithography." In *Five Centuries of Map Printing,* edited by David Woodward, 137–55. Chicago: University of Chicago Press, 1975.

Korea Times. "US University Library Lists Old Maps of Korea's East Sea on its Website," October 9, 2002.

Korean Overseas Information Service. "Rediscover the Proper Name for Korea's East Sea." Issue statement, October 2002. http://www.korea.net/issue/eastsea/back.html.

Kors, Alan Charles, and Harvey A. Silverglate. *The Shadow University: The Betrayal of Liberty on America's Campuses.* New York: Free Press, 1998.

Kross, John F. "What's in a Name?" *Sky and Telescope* 89 (May 1995): 28–33.

Ladbury, Sarah, and Russell King. "Settlement Renaming in Turkish Cyprus." *Geography* 73 (1988): 363–67.

Ladwig, Alan. "When Cape Crusaders Played Florida's Name Game." SPACE.com history page, May 13, 2000. http://www.space.com/news/spacehistory/cape_kennedy_000511.html.

Lagerfeld, Steven. "Name That Dune." *Atlantic Monthly,* September 1990.

Large, Arlen J. "Naughty Girl Meadow by Some Other Name Is Known and Loved." *Wall Street Journal,* April 3, 1972.

Lawrence, David. *Upheaval from the Abyss: Ocean Floor Mapping and the Earth Science Revolution.* New Brunswick, NJ: Rutgers University Press, 2002.

Ledyard, Gari. "Cartography in Korea." In *The History of Cartography.* Vol. 2, book 2, *Cartography in the Traditional Asian and Southeast Asian Societies,* edited by J. B. Harley and David Woodward, 235–345. Chicago: University of Chicago Press, 1994.

Lee, Ki-suk. "East Sea in World Maps." Issue statement, Korean Overseas Information Service. http://www.korea.net/issue/eastsea/map_foreword.asp.

Lerch, Lothar. "Many VT-er's Most Desired Travel Destination." Lotharlerch's Fucking Travel Page. VirtualTourist.com. http://www.virtualtourist.com/m/108df/50a96.

Loewen, James W. *Lies Across America: What Our Historic Sites Get Wrong.* New York: New Press, 2000.

Louis, Renee. "The Authoritative Textualization of Hawaiian Place Names." Master's thesis, University of Hawaii, 1998.

———. "Indigenous Hawaiian Cartographer: In Search of Common Ground." *Cartographic Perspectives,* no. 48 (Spring 2004): 7–23.

Lynch, Marika. "Fishy Nickname Officially Abandoned." *San Diego Union-Tribune,* May 30, 2001.

MacDonald, Burton. *East of the Jordan: Territories and Sites of the Hebrew Scriptures.* Boston: American Schools of Oriental Research, 2000.

Maine Legislature. "An Act Concerning Offensive Names." 119th session, legislative documents LD 2419 and LR 3466. http://www.mainelegislature.org/legis/bills_119th/.

Maines, Rachel P. *The Technology of Orgasm: "Hysteria," the Vibrator, and Women's Sexual Satisfaction.* Baltimore: Johns Hopkins University Press, 1998.

Marshall Penn-York Co. *Visual Encyclopedia® Fully Indexed Street Map of Metropolitan Syracuse and Onondaga County, New York.* Syracuse, NY: Marshall Penn-York, 1976, revised 1979.

———. *Visual Encyclopedia® Map of Syracuse, N.Y.* Syracuse, NY: Marshall Penn-York, 1997.

Mattson, Mark. "Imagesetting in Desktop Mapping." *Cartographic Perspectives,* no. 6 (Summer 1990): 13–22.

McArthur, Lewis L. "The GNIS and the PC: Two Tools for Today's Toponymic Research." *Names* 43 (1995): 245–54.

———. *Oregon Geographic Names.* 4th ed., rev. and enlarged. Portland: Oregon Historical Society, 1974.

McCarthy, Michael. "Geography: We'd Be Lost without It." *Independent* (London), August 19, 2004.

Meleagrou, Eleni, and Birol Yesilada. "The Society and Its Environment." In *Cyprus: A Country Study,* edited by Eric Solsten, 47–103. Library of Congress Area Handbook Series. Washington, DC: U.S. Government Printing Office, 1993.

Melmer, David. "Offensive Names, Nursing Homes among South Dakota Bills." *Indian Country Today,* February 5, 2001.

Merriam, Hart (chairman), and Jas. McCormick (secretary). *Report of the United States Geographic Board on S. J. Res. 64, a Joint Resolution to Change the Name of "Mount Rainier" to "Mount Tacoma" and for Other Purposes.* Washington, DC: U.S. Government Printing Office, 1924.

Merriam-Webster's Geographical Dictionary. 3rd ed. Springfield, MA: Merriam-Webster, 1997.

Microsoft Corporation. "Microsoft's Geopolitical Product Strategy Team Helps Prevent Worldwide Geographical, Political and Cultural Issues from Becoming International Incidents." Press release, December 6, 2000. http://www.microsoft.com/presspass/features/2000/Dec00/12–06gps.asp.

MidEast Web Maps. "The Israeli Camp David II Proposals for Final Settlement." http://www.mideastweb.org/campdavid2.htm.

———. Palestine maps. http://www.mideastweb.org/maps.htm.

Miller, Keith. "It's All in a Name." *Geographical* 68 (June 1996): 10.

Miller, Larry L. *Ohio Place Names.* Bloomington: Indiana University Press, 1996.

Miller, O. M. "Planetabling from the Air: An Approximate Method of Plotting from Oblique Aerial Photographs." *Geographical Review* 21 (1931): 201–12.

Ministry of Foreign Affairs of Japan. "Sea of Japan," August 2002. http://www.mofa.go.jp/.

Minnesota State Senate. S.F. No. 574, 2nd Engrossment, 79th Legislative Session (1995–96). http://www.revisor.leg.state.mn.us.

Minority Rights Group. *Cyprus*, report no. 30. London, 1976.

Minton, Arthur. "Names of Real-Estate Developments." *Names* 7 (1959): 129–53, 233–55.

Monahan, David. "The General Bathymetric Chart of the Oceans: Seventy-three Years of International Cooperation in Small Scale Thematic Mapping." *International Hydrographic Review* 54 (July 1977): 9–18.

Monmonier, Mark. *Drawing the Line: Tales of Maps and Cartocontroversy*. New York: Henry Holt, 1995.

———. "Map Traps: the Changing Landscape of Cartographic Copyright." *Mercator's World* 6 (July/August 2001): 50–52.

Moore, Terris. *Mt. McKinley: The Pioneer Climbs*. College: University of Alaska Press, 1967.

Morehead, Joe. "Decision Lists and the U.S. Board on Geographic Names." *Serials Librarian* 5 (Spring 1981): 7–13.

Morris, Allen. *Florida Place Names*. Coral Gables, FL: University of Miami Press, 1974.

Morris, Benny. *The Birth of the Palestinian Refugee Problem, 1947–1949*. Cambridge: Cambridge University Press, 1987.

———. *1948 and After: Israel and the Palestinians*. Oxford: Clarendon Press, 1990.

Muldrow, Robert. "Mount McKinley." *National Geographic* 12 (1901): 312–13.

Müller-Wille, Ludger. "Inuit Toponymy and Cultural Sovereignty." In *Conflict in Development in Nouveau-Québec*. McGill Subarctic Research Paper no. 37, edited by Ludger Müller-Wille, 131–50. Montreal: McGill University Centre for Northern Studies and Research, 1983.

Murray, John. *Politics and Place-Names: Changing Names in the Late Soviet Period*. Birmingham Slavonic Monographs no. 32. Birmingham, UK: University of Birmingham, Department of Russian, 2000.

Nainan, Madhu. "'Mumbai' to Replace Bombay on Map." Agence French Presse News Wire, April 30, 1995.

Name A Star, Inc. "Virtual Star Package." http://www.nameastarinc.com/.

National Geographic Atlas of the World. 7th ed. Washington, DC: National Geographic Society, 1999.

National Geographic Society. "Keep Your Atlas Up to Date." Atlas Updates. http://www.nationalgeographic.com/maps/atlas_updates.html.

———. "Sea of Japan." Atlas Updates, high-resolution patch, 2001. http://www.nationalgeographic.com/maps/pdfs/seaofjapan.pdf.

———. "Sea of Japan (East Sea)." Atlas Updates, 2001. http://www.nationalgeographic.com/maps/updates/seaofjapan.html.

National Geospatial-Intelligence Agency. GEOnet Names Server. http://earth-info.nga.mil/gns/html/index.html.

National Imagery and Mapping Agency. GEOnet Names Server (GNS). http://earth-info.nima.mil/gns/html/index.

National Oceanic and Atmospheric Administration, National Marine Fisheries Service, Office of Protected Resources. "Goliath Grouper (*Epinephelus itajara*)." http://www.nmfs.noaa.gov/prot_res/species/fish/goliath_grouper.html.

National Oceanic and Atmospheric Administration, World Data Center for Marine Geology and Geophysics. "History of GEBCO, 1903–2003: Book Reviews." http://www.ngdc.noaa.gov/mgg/gebco/bookreviews.html.

National Park Service. "Denali National Park and Preserve Mountaineering Booklet." http://www.nps.gov/dena/home/mountaineering/booklet/mbenglish.html.

———. "Denali National Park and Preserve Mountaineering History." http://www.nps.gov/dena/home/mountaineering/history/hx.html.

National Research Council, Mapping Science Committee, Committee to Review the U.S. Geological Survey Concept of the National Map. *Weaving a National Map: Review of the U.S. Geological Survey Concept of the National Map.* Washington, DC: National Academies Press, 2003.

Natural Resources Canada. Atlas of Canada, "Nunavut, 2001," reference map. http://atlas.gc.ca/rasterimages/english/maps/reference/provincesterritories/nunavut.pdf.

———. Canadian Geographical Names. http://geonames.nrcan.gc.ca/info/princip1990_e.php.

"The New Europe." *National Geographic* 182 (December 1992), map supplement.

"New Maps of the Antarctic." *Geographical Review* 30 (1940): 160–62.

"New Undersea Features." *International Hydrographic Bulletin,* September 1966, 304.

New York State Committee on Geographic Place Names. Policies and Procedures (draft), November 3, 1989.

New York Times. "Burma Out, Myanmar In," June 25, 1989.

———. "Corrections," November 11, 1986.

———. "14 Arrested in Protest," December 11, 1984.

———. "Politics Intrude on Street Name Hearing," November 1, 1986.

———. "What's in a Name? Alaska Battles Ohio over Title for Mt. McKinley," November 16, 1980.

Newson, D. W. "The General Bathymetric Chart of the Oceans—Seventy Years of International Cartographic Co-operation." *Cartographic Journal* 8 (1971): 39–47.

Nicolaisen, W. F. H. "Placenames and Politics." *Names* 38 (1990): 193–207.

Nieminen, Anna. "The Cultural Politics of Place Naming in Québec: Toponymic Negotiation and Struggle in Aboriginal Territories." PhD diss., University of Ottawa, 1998.

Nijhuis, Michelle. "Tribal Immersion Schools Rescue Language and Culture." *Christian Science Monitor,* June 11, 2002.

"Nomenclature of Ocean Bottom Features." *International Hydrographic Review* 48 (January 1971): 203–8.

Norris, Darrell A. "Unreal Estate: Words, Names and Allusions in Suburban Home Advertising." *Names* 47 (1999): 365–80.

North Carolina Legislature. Information/history, General Assembly Bill 483, 2003–2004 sess. http://www.ncleg.net/homePage.pl.

"Ocean Floor Discoveries in the Atlantic." *International Hydrographic Bulletin,* November 1970, 386–88.

O'Dwyer, Thomas. "Candle in the Sea." *Jerusalem Post,* July 15, 1997.

Okimoto, Jolyn. "Board Rejects 'Phoenix Peak' as Alternative Name for Skyline Landmark." Associated Press, January 20, 1999.

Oregon Legislature. Senate Bill 488, signed by the governor on June 27, 2001. http://www.leg.state.or.us/orlaws/sess0600.dir/0652ses.html.

———. Senate Joint Memorial no. 3, signed June 27, 2001. http://www.leg.state.or.us/99orlaws/sessresmem.dir/sjm0003.html.

Ormeling, Ferjan. "Minority Toponyms—the Western European Experience." *ITC Journal,* no. 1 (1984): 58–63.

Orth, Donald J. "The Mountain Was Wronged: The Story of the Naming of Mt. Rainier and Other Domestic Names Activities of the US Board on Geographic Names." *Names* 32 (1984): 428–34.

———. *Principles, Policies, and Procedures: Domestic Geographic Names.* Reston, VA: U.S. Department of the Interior, 1989.

———. "The U.S. Board on Geographic Names: An Overview." *Names* 38 (1990): 165–72.

Orth, Donald J., and Elizabeth Unger Mangan. *Geographic Names and the Federal Government: A Bibliography.* Washington, DC: Library of Congress, Geography and Map Division, 1990.

Orth, Donald J., and Roger L. Payne. *The National Geographic Names Data Base: Phase II Instructions.* U.S. Geological Survey Circular 1011. Washington, DC: U.S. Government Printing Office, 1987.

———. *Principles, Policies, and Procedures: Domestic Geographic Names.* Washington, DC: U.S. Board on Geographic Names, 1997.

———. *Principles, Policies, and Procedures: Domestic Geographic Names.* Online edition, revised, 2003. http://geonames.usgs.gov/pppdgn.html.

———. *Principles, Policies, and Procedures: Domestic Geographic Names.* Print version, revised. Washington, DC: U.S. Board on Geographic Names, 2003.

Papadopoulos, Tassos (president of Republic of Cyprus). Speech at the official dinner in his honour by the President of the Republic of Greece, November 25, 2003. Press release, Republic of Cyprus Press and Information Office. http://www.pio.gov.cy/news/special_issues/special_issue152.htm.

Papenfuse, Edward C., and others. *Maryland: A New Guide to the Old Line State.* Baltimore: Johns Hopkins University Press, 1976.

Parry, R. B., and C. R. Perkins. *World Mapping Today.* London: Butterworths, 1987.

Payne, Roger L. "Applied Toponymy." *Names* 49 (2001): 293–99.

———. "Development and Implementation of the National Geographic Names Database." *Names* 43 (1995): 307–14.

———. "Geographic Names Information System: Philosophy and Function." *World Cartography* 18 (1986): 49–52.

————. "The United States Board on Geographic Names: Standardization or Regulation?" *Names* 48 (2000): 177–92.

Peel, Quentin. "Trouble Is Brewing for Europe over Cyprus." *Financial Times* (London), July 8, 2004.

Peritz, Ingrid. "Quebec Names Islands to Mark Bill 101." *Ottawa Citizen*, August 23, 1997.

Permanent Committee on Geographical Names for Official British Use. http://www.pcgn.org.uk/Indexx.htm.

Philip's Concise World Atlas. 10th ed. London: George Philip Ltd., 2000.

Phillips, Hugh A. "Copperplate Engraving for the Production of Topographic Maps at the United States Geological Survey, 1890–1953." *Meridian* [Map and Geography Round Table, American Library Association] 11 (1997): 5–21.

Phoenix New Times. "Squaw Pique: Cast Your Vote in the New Times Totem Poll," December 30, 1992.

Pierre, Robert E. "Northwest Tribe Struggles to Revive Its Language." *Washington Post*, March 31, 2003.

Pizer, Vernon. "The Place Namers." *American Legion Magazine* 108 (January 1980): 16–17, 44–47.

Population Reference Bureau. DataFinder. http://www.prb.org/datafind/datafinder.htm.

Powell, William Stevens. *The North Carolina Gazetteer*. Chapel Hill: University of North Carolina Press, 1968.

Proctor, A. *Other Worlds than Ours*. New York: P. F. Collier and Son, 1902.

Programma Nazionale di Ricerche in Antartide. Composite Gazetteer of Antartica: History of the Work Programme. http://www3.pnra.it/LUOGHI_ANT/HTML_en/history.html.

Pukui, Mary Kawena, Samuel H. Elbert, and Esther T. Mookini. *Place Names of Hawaii*. 2nd ed. Honolulu: University Press of Hawaii, 1974.

Pyati, Archana. "Legislators Vote to Cut 'Squaw' from Place Names." *Portland Oregonian*, June 1, 2001.

Quindlen, Anna. "A Day for Renaming Places." *New York Times*, April 23, 1983.

Randall, Richard R. *Place Names: How They Define the World—and More*. Lanham, MD: Scarecrow Press, 2001.

————. "The U.S. Board on Geographic Names and International Programs." *Names* 39 (1991): 249–56.

————. "The U.S. Board on Geographic Names and Its Work in Foreign Areas." *Names* 38 (1990): 173–82.

————. "What's in a (Place) Name? United Nations and Other Programs on Geographic Names: How U.S. Geographers Have Played a Role." *Revista Cartografica* 63 (1995): 69–82.

Raup, Hallock F. "The Names of Ohio's Streams." *Names* 5 (1957): 162–68.

Rayburn, Alan. *Naming Canada: Stories about Canadian Place Names*. Revised and expanded. Toronto: University of Toronto Press, 2001.

————. "Native Names for Native Places." *Canadian Geographic* 107 (April/May 1987): 88–89.

———. "Unfortunate Connotations Acquired by Some Canadian Toponyms." *Names* 36 (1988): 187–92.

Reeder, Jim. "Bills Target Racially Sensitive Place Names." *Palm Beach Post*, March 29, 2004.

Reese, Gary Fuller. *Pierce County Name Origins with Origins of the Name Tacoma.* Tacoma, WA: Tacoma Public Library, 1978.

Reid, T. R. "S. Koreans Almost Sink Sea of Japan Plan." *Washington Post*, September 24, 1994.

René, Nicole. "Op-Ed: Current Use Determines Name." *Montreal Gazette*, April 7, 1998.

Rennick, Robert M. "On the Success of Efforts to Retain the Names of Several American Communities in the Two World Wars." *Names* 32 (1984): 26–32.

Republic of Cyprus Press and Information Office. "Results of Invasion." Press release, January 15, 2004. http://www.moi.gov.cy/moi/PIO/PIO.nsf/All/2BAA147846E3D0D8C2256D6D001E84F8?OpenDocument.

Republic of Turkey, Ministry of Tourism. "Turkey Tourist Map," 1:1,850,000 [explanations in English]. Istanbul, 1994.

Rischbach, Michael R. "An Answer for Jews and Palestinians." *Cleveland Plain Dealer*, May 7, 2003.

Robinson, Arthur H., and Randall D. Sale. *Elements of Cartography.* 3rd ed. New York: John Wiley and Sons, 1969.

Robinson, Greg. *By Order of the President: FDR and the Internment of Japanese Americans.* Cambridge, MA: Harvard University Press, 2001.

Rodriguez, Gregory. "The Soul of a New Neighborhood." *Los Angeles*, January 1997.

Roig-Franzia, Manuel. "Coast to Coast." *Washington Post*, December 8, 2002.

Romig, Walter. *Michigan Place Names: The History of the Founding and the Naming of More than Five Thousand Past and Present Michigan Communities.* Detroit: Wayne State University Press, 1986.

Rosario, Sonya. "WOCA Builds Power for Women of Color in Idaho." *@tr — News and Events for the Progressive Movement* [newsletter of A Territorial Resource], April 2003. http://www.atrfoundation.org/news/2003_04/woca.htm.

Rosenberg, Matt. "Sea of Japan vs. East Sea: Letter Writing Campaign Influences Cartography." About Geography, February 24, 2002. http://geography.about.com/library/weekly/aa022402a.htm.

Roslin, Alex, and Ernest Webb. "Iles 101: Quebec Admits Mistake." *Montreal Gazette*, September 18, 1997.

Rundstrom, Robert A. "A Cultural Interpretation of Inuit Map Accuracy." *Geographical Review* 80 (1990): 155–68.

———. "The Role of Ethics, Mapping, and the Meaning of Place in Relations between Indians and Whites in the United States." *Cartographica* 30 (Spring 1993): 21–28.

Rutherford, Phillip R. "Censorship and Some Maine Place Names." In *Of Edsels and Marauders*, edited by Fred Tarpley and Ann Moseley, 47–50. Commerce, TX: South Central Names Institute, 1971.

Sachs, Susan. "With Eye on Europe, Turkey Backs Bid to Reunite Cyprus." *New York Times,* April 21, 2004.

Safire, William. "Bring Back Upper Volta." *New York Times Magazine,* October 22, 1989.

Sagan, Carl. *Billions and Billions: Thoughts on Life and Death at the Brink of the Millennium.* Thorndike, ME: Thorndike Press, 1998.

Sahlberg, Bert. "What's in a Name? In This Case, Too Many Letters." *Lewiston (ID) Morning Tribune,* August 9, 2002.

St. Georges High School, Dildo, Newfoundland. K–12 Internet On Ramp, Newfoundland Educational Networking Group. "Dildo Days." http://www.k12.nf.ca/stgeorgeshigh/Dildo/DildoDays.htm.

San Miguel, George L. "How Is Devils Tower a Sacred Site to American Indians?" National Park Service, Devils Tower National Monument, August 1994. http://www.nps.gov/deto/place.htm.

Sandford, H. A. "The Geographical Name in Modern School Atlases: A Study of the Reasons for the Persistence of Exonyms Despite the Increasing Worldwide Adoption of Standardised Names." *Cartographic Journal* 27 (1990): 137–41.

Sargeant, Frank. "Name Change Not Exactly a 'Goliath' Issue." *Tampa Tribune,* May 27, 2001.

Schanberg, Sydney H. "Jackson as Polarizer." *New York Times,* April 10, 1984.

Schmitt, Eric. "Lake County Journal: Battle Rages over a 5-letter Four-letter Word." *New York Times,* September 4, 1996.

———. "North Korea MIG's Intercept U.S. Jet on Spying Mission." *New York Times,* March 4, 2003.

———. "Ultimate Arbiter of Hill and Vale." *New York Times,* November 27, 1986.

Schwartz, John. "Speaking Out and Saving Sounds to Keep Native Tongues Alive." *Washington Post,* March 14, 1994.

Shaffer, David E. "Korea, Corea, Gorea, or Koria?" *Korea Herald,* August 30, 2003.

Shaw, David L. "Tubman Dedication Set for May 27." *Syracuse Post-Standard,* January 10, 2004.

Shuger, Scott. "Los Angeles Postcard: Name Game." *New Republic,* July 29, 1991.

Siegel, Lee. "Name Changes for Landmarks Sign of the Times." *Salt Lake Tribune,* September 7, 1996.

Sievers, Jörn, and Peter Jordan. "International Workshop on Exonyms 'GeoNames 2001,' Berchtesgaden, 1–2 October 2001." *Bundesamt für Kartographie und Geodäsie,* June 14, 2002. http://www.ifag.de/kartographie/Stagn/GeoNames2001.htm.

Smith, Christopher. "Tribes Say Devils Tower Is No Name for a Pious Peak." *Salt Lake Tribune,* September 4, 1996.

Solsten, Eric, ed. *Cyprus: A Country Study.* Library of Congress Area Handbook Series. Washington, DC: U.S. Government Printing Office, 1993.

South Dakota Legislature. "Geographic Place Names Replaced." 2001 Session Laws, chap. 9. http://legis.state.sd.us/sessions/2001/sesslaws/ch009.htm.

Steinwehr, Adolph Wilhelm August Friedrich von. *The Centennial Gazetteer of the United States.* Philadelphia: J. C. McCurdy and Co., 1873.

Stewart, George Rippey. *American Place-Names: A Concise and Selective Dictionary for the Continental United States of America*. New York: Oxford University Press, 1970.

———. "A Classification of Place Names." *Names* 2 (March 1954): 1–13.

———. *Names on the Globe*. New York: Oxford University Press, 1975.

———. *Names on the Land*. Boston: Random House, 1945.

Stokke, Olav Schram. "The Making of Norwegian Antarctic Policy." In *Governing the Antarctic: The Effectiveness and Legitimacy of the Antarctic Treaty System*, edited by Olav Schram Stokke and Davor Vidas, 384–408. Cambridge: Cambridge University Press, 1996.

Stokke, Olav Schram, and Davor Vidas, eds. *Governing the Antarctic: The Effectiveness and Legitimacy of the Antarctic Treaty System*. Cambridge: Cambridge University Press, 1996.

Streeter, Dawn-Marie. "Pressing for Change of Venerable Names." *New York Times*, February 26, 1995.

Strobell, Mary E., and Harold Masursky. "Planetary Nomenclature." In *Planetary Mapping*, edited by Ronald Greeley and Raymond M. Batson, 96–140. Cambridge: Cambridge University Press, 1990.

Stuck, Hudson. *The Ascent of Denali: A Narrative of the First Complete Ascent of the Highest Peak in North America*. New York: Charles Scribner' Sons, 1914. Reprint, Lincoln: University of Nebraska Press, 1989.

Sullivan, Walter. "Honoring, and Unearthing, Indian Place Names." *New York Times*, September 9, 1990.

Sunday Tribune (Ireland), "Fucking Villagers Vote against Name Change," June 13, 2004.

"A Talk with Jesse Jackson." *Newsweek*, April 9, 1984.

Tartter, Jean R. "National Security." In *Cyprus: A Country Study*, edited by Eric Solsten, 211–43. Library of Congress Area Handbook Series. Washington, DC: U.S. Government Printing Office, 1993.

Taule, Corey. "Protestors of 'Squaw' Get No Answers from Kempthorne." *Idaho Falls Post Register*, March 10, 2001.

———. "'The Ugliest Word I've Ever Heard'—Process Begins to Remove 'Squaw' from Idaho Geography." *Idaho Falls Post Register*, February 2, 2001.

Thomas, Brad. "US Policy on Name of Burma." *Geographic Notes* [U.S. Department of State], no. 11 (March 1, 1990): 12.

Thorne, Tony. *The Dictionary of Contemporary Slang*. New York: Pantheon Books, 1990.

Thrower, Norman J. W. *Original Survey and Land Subdivision: A Comparative Study of the Form and Effect of Contrasting Cadastral Surveys*. Chicago: Rand McNally for the Association of American Geographers, 1966.

Toronto Star. "Korea Out to Rename the Sea of Japan," February 16, 1993.

———. "No Safe Middle Ground in the Map Wars," August 28, 1999.

Trimble, Virginia. "What, and Why, Is the International Astronomical Union?" *Beam Line* [Stanford Linear Accelerator] 27 (Winter 1997): 43–52.

Turco, Peggy. *Walks and Rambles in Dutchess and Putnam Counties*. Woodstock, VT: Countryman Press, 1990.

Turkish Republic of Northern Cyprus, Deputy Prime Minister and Ministry of Foreign Affairs, Public Relations Department. "Introductory Survey." http://www.trncinfo.com/TANITMADAIRESI/2002/ENGLISH/ALLaboutTRNC/Page02.htm.

Tyson, Ann Scott. "Why Peking Becomes Beijing." *Christian Science Monitor,* January 23, 1989.

"Ugh! Oops." *New Republic,* February 18, 1991.

"Uncharted Undersea Mountains, Mountain Ridges and Sea Basins Discovered in North Pacific." *International Hydrographic Bulletin,* March 1967, 86–88.

United Nations Department of Economic and Social Affairs, Statistics Division. "Geographical Names: Overview." http://unstats.un.org/unsd/geoinfo/about_us.htm.

———. "Geographical Names: UNGEGN Working Groups." http://unstats.un.org/unsd/geoinfo/ungegnwgroups.htm.

United Nations General Assembly. "Resolution 37/253, adopted May 16, 1983." http://www.un.org/documents/ga/res/37/a37r253.htm.

United Nations Group of Experts on Geographical Names. *Glossary of Terms for the Standardization of Geographical Names.* UN document ST/ESA/STAT/SER.M/85. New York: United Nations, 2002.

———. "Working Group on Exonyms of the United Nations Group of Experts on Geographical Names." Eighth United Nations Conference on the Standardization of Geographical Names, Berlin, 27 August–5 September 2002. UN document E/CONF.94/3, 36–37.

United Nations Sub-Commission on the Prevention of Discrimination. "Violations of Human Rights in Cyprus, September 2, 1987." Resolution 987/19. Republic of Cyprus Press and Information Office. http://www.moi.gov.cy/moi/PIO/PIO.nsf/All/EDB081B5966DD444C2256DC100364BDB/$file/Resolution%20of%20the%20Subcommission.doc?OpenElement.

Urbanek, Mae Bobb. *Wyoming Place Names.* Boulder, CO: Johnson Publishing, 1975. Reprint, Missoula, MT: Mountain Press, 1988.

U.S. Board on Geographic Names. *Decisions on Geographic Names in the United States: Decision List 1997.* Washington, DC: Department of the Interior, 1997.

———. *Decisions on Geographic Names in the United States: Decision List 1998.* Washington, DC: Department of the Interior, 1998.

———. *Decisions on Geographic Names in the United States: Decision List 1999.* Washington, DC: Department of the Interior, 1999.

———. *Decisions on Geographic Names in the United States: Decision List no. 8902* (July–September 1989). Washington, DC: Department of the Interior, 1989.

———. "Domestic Geographic Name Proposal." Online submission form. http://geonames.usgs.gov/dgnp/dgnp.html.

———. *Field Investigation of Native American Place Names.* Washington, DC, [1990?].

———. *First Report on Foreign Geographic Names, 1932.* Washington, DC: U.S. Government Printing Office, 1932.

———. *First Report of the United States Board on Geographic Names, 1890–1891.* Washington, DC: U.S. Government Printing Office, 1892.

———. *Fourth Report of the United States Geographic Board, 1890–1916.* Washington, DC: U.S. Government Printing Office, 1916.

———. *Policies and Guidelines for the Standardization of Undersea Feature Names* (draft approved April 6, 1999). http://earth-info.nga.mil/gns/html/acuf/GUIDE99.pdf.

U.S. Board on Geographical Names. *Decisions of the United States Board on Geographical Names: Decisions Rendered between July 1, 1938 and June 30, 1939.* Washington, DC: U.S. Government Printing Office, 1939.

———. *The Geographical Names of Antarctica.* Special Publication no. 86. Washington, DC: U.S. Government Printing Office, 1947.

U.S. Central Intelligence Agency. "Antarctic Region," ca. 1:45,000,000, map no. 802917AI (R02207) 6–02. Washington, DC, 2002.

———. "Cyprus," World Factbook. http://www.cia.gov/cia/publications/factbook/geos/ks.html.

———. "The Disputed Area of Kashmir," ca. 1:5,100,000, map no. 760279AI (R00744) 6–02. Washington, DC, 2002.

———. *The Gaza Strip and West Bank—A Map Folio.* Washington, DC, 1994.

———. "Korea, South," World Factbook. http://www.cia.gov/cia/publications/factbook/geos/ks.html.

U.S. Congress. HR 164, 108th Cong., 1st sess. http://thomas.loc.gov.

———. Wilderness Act of 1964. P.L. 88–577. http://www.leaveitwild.org/reports/wilderness1964PF.html.

U.S. Department of State. "Background Note: Burma," December 2003. http://www.state.gov/r/pa/ei/bgn/19147.htm.

———. "Guidance Bulletin No. 12, February 18, 1994: Guidance on Depiction of the Former Yugoslav Republic of Macedonia." *Geographic and Global Issues Quarterly* 3, no. 4 (Winter 1994): 18–19.

U.S. Geographic Board. *Sixth Report of the United States Geographic Board, 1890 to 1932.* Washington, DC: U.S. Government Printing Office, 1933.

U.S. Geological Survey. "Automated Cartographic Lettering." In *Geological Survey Research, Fiscal Year 1981.* U.S. Geological Survey Professional Paper 1375, 307–8. Washington, DC: U.S. Government Printing Office, 1984.

———. "Frequently Asked Questions about GNIS." http://geonames.usgs.gov/faqs.html#7.

———. *Geographic Names Information System Data Users Guide 6.* 4th printing, revised. Reston, VA: USGS, 1995. http://geonames.usgs.gov/gnis_users_guide_appendixc.html.

———. GNIS Antarctic Query Web page. http://geonames.usgs.gov/antform.html.

———. *National Gazetteer of the United States of America, United States Concise.* U.S. Geological Survey Professional Paper 1200-US. Washington, DC: U.S. Government Printing Office, 1990.

————. *National Gazetteer of the United States of America—Delaware, 1983.* U.S. Geological Survey Professional Paper 1200-DE. Washington, DC: U.S. Government Printing Office, 1984.

————. *National Gazetteer of the United States of America—New Jersey, 1983.* U.S. Geological Survey Professional Paper 1200-NJ. Washington, DC: U.S. Government Printing Office, 1983.

————. "Policy Covering Antarctic Names." http://geonames.usgs.gov/antex .html.

————. *United States Geological Survey Annual Report, Fiscal Year 1975.* Washington, DC: U.S. Government Printing Office, 1976.

————. *The United States Geological Survey: Its Origin, Development, Organization, and Operations.* U.S. Geological Survey Bulletin 227. Washington, DC: U.S. Government Printing Office, 1904.

U.S. Geological Survey, Astrogeology Research Program. "Eros Nomenclature: Crater." Gazetteer of Planetary Nomenclature. http://planetarynames.wr.usgs .gov/asteroids/eroscrat.html.

————. Gazetteer of Planetary Nomenclature. http://planetarynames.wr.usgs .gov/.

————. "How Names Are Approved." Gazetteer of Planetary Nomenclature. http://planetarynames.wr.usgs.gov/approved.html.

————. "IAU Rules and Conventions." Gazetteer of Planetary Nomenclature. http://planetarynames.wr.usgs.gov/rules.html.

————. "IAU Working Group and Task Group Members." Gazetteer of Planetary Nomenclature. http://planetarynames.wr.usgs.gov/append2.html (accessed May 17, 2004).

U.S. Geological Survey, Hawaiian Volcano Observatory. "Summary of the Pu'u 'Ō'o–Kupaianaha Eruption, 1983–Present." http://hvo.wr.usgs.gov/kilauea/ summary/.

U.S. Geological Survey, Office of Geographic Names. "Geographic Names Information System, National Geographic Names Data Compilation Program, July 2002." Status map, June 25, 2003. http://geonames.usgs.gov/statusmap.html.

U.S. Library of Congress. *A World of Names: Celebrating the Centennial, United States Board on Geographic Names.* Washington, DC: Library of Congress, 1990.

U.S. Navy, Arctic Submarine Laboratory. "Historical Timeline." http://www.csp .navy.mil/asl/Timeline.htm.

U.S. Navy, Naval Meteorological and Oceanographic Command. "Landry Bernard Retires as NAVO Technical Director." News Online, press release, July/August 2001. http://pao.cnmoc.navy.mil/pao/n_online/archive/vol21no4/articles/ bernard.htm

U.S. Office of Geography. *Antarctica: Official Standard Name Decisions of the United States Board on Geographic Names.* Gazetteer no. 14. Washington, DC: U.S. Government Printing Office, 1966.

Vasiliev, Irina Ren. *From Abbotts to Zurich: New York State Placenames.* Syracuse, NY: Syracuse University Press, 2004.

————. "Mapping Names." *Names* 43 (1995): 294–306.

————. "The Naming and Diffusion of Moscows across the United States." Master's thesis, State University of New York at Buffalo, 1988.

————. "The Naming of Moscows in the USA." *Names* 37 (1989): 51–64.

Verner, Coolie. "Copperplate Engraving." In *Five Centuries of Map Printing*, edited by David Woodward, 51–75. Chicago: University of Chicago Press, 1975.

Vogel, Virgil J. *Indian Names in Michigan*. Ann Arbor: University of Michigan Press, 1986.

Washington Post. "Negro Mountain Keeps Name," February 2, 1995, Maryland Weekly sec.

Washington (D.C.) Star-News. "People in the News," December 5, 1972.

Whitaker, Ewen A. *Mapping and Naming the Moon: A History of Lunar Cartography and Nomenclature*. Cambridge: Cambridge University Press, 1999.

Whitbeck, R. H. "Geographic Names in the United States and the Stories They Tell." *National Geographic* 16 (1905): 100–104.

White, Robert C. "National Gazetteers of the United States." *Names* 18 (1970): 9–19.

Whittlesey, Lee H. *Yellowstone Place Names*. Helena: Montana Historical Society Press, 1988.

Willing, Richard. "'Cripple Creek,' 'Squaw Tits,' and Other Mapmaking No-Nos." *Washingtonian*, June 1996.

Wood, Houston. *Displacing Natives: The Rhetorical Production of Hawai'i*. Lanham, MD: Rowman and Littlefield, 1999.

World Atlas of Nations. Chicago: Rand McNally, 1988.

Wraight, A. J. "Field Work in the U.S.C.&G.S." *Names* 2 (1954): 153–62.

Writers' Program, Works Progress Administration. *Pennsylvania: A Guide to the Keystone State*. New York: Oxford University Press, 1940.

WuDunn, Sheryl. "Koreans, Still Bitter, Recall 1945." *New York Times*, May 2, 1995.

Zelinsky, Wilbur. "Along the Frontiers of Name Geography." *Professional Geographer* 49 (1997): 465–66.

————. "The Game of the Name." *American Demographics* 11 (April 1989): 42–45.

Index

Devils Tower, WY, 72–73
Devils Tower National Monument, 73
diacritical marks, 82–83, 97, 134,
165n32
Dickey, William A., 74–75
Dildo, Newfoundland, 67–69
dingleberries, 69
Dingle Hole Wildlife Management
Area, NY, 69–70
Dingman, Lester, 61
Dogshit Park, CA, 70–71
Dole, Sanford, 81
Doleman, Marguerite, 44
Dudley, Robert, 91
duplicate or multiple toponyms: fed-
eral policy against, 11, 36, 147; in
multilingual countries, 97; as
potential source of confusion, 11,
91, 97; of streets, 122–23

East Palo Alto, CA, 47
East Sea, proposed as replacement
name for Sea of Japan, 90–95, 148,
169n15
Elford, Robert, 67
Encyclopaedia Britannica, online version, 92
endonyms, 99
English Atlas, The (1711), 91
English speakers, toponyms offensive
to, 39
environmentally offensive toponyms,
71
Eros (asteroid), toponyms for, 141
Eskimo. *See* Inuit
exonyms, 99–100, 110

feature names, as distinct from place
names, 8–9. *See also* toponyms
feature separates, 19, 154n6
feature types, 2, 136–38
Federal Aviation Administration
(FAA), 32
Federal Communications Commission (FCC), 32
Feist v. Rural Telephone Company, 127

First List of Names in Palestine, 113
fishermen, as source of offensive topo-
nyms, 62
Fish Lake, OR, 65
Florida: federal names list issued for,
28; resistance to renaming Cape
Canaveral, 5–6
folk-etymology toponyms, 8
Ford, Jeff, 57
Foreign Names Information Bulletin, 96,
170n33
Fram Strait, 136, 137
French Lick, IN, 61
Frobisher Bay (old name for Iqaluit),
Nunavut, 88
Fucking (Austrian village), 162n18

Gannett, Henry, 22, 154n16, 154n18
Gayside, Newfoundland, 67
Gaza Strip, 102
Gazetteer of Colorado, A (1906), 22
Gazetteer of Israel (compiled by U.S.
Defense Mapping Agency), 116
gazetteers, 21, 154n16; on CD-ROM,
29; foreign names, 96–97;
national, 21–26, 29; state, 22–23,
28–29, 155n24
GEBCO Digital Atlas, 135
General Bathymetric Chart of the
Oceans (GEBCO), 134, 135–36
generic (element of a toponym), 6–8,
57, 79
Geographical (magazine), 104
geographical dictionaries, 21, 154n16
geographic information system (GIS),
20
geographic names. *See* toponyms
Geographic Names Information
System (GNIS), x, 30, 146; Antarctic features in, 131, 132, 175n32;
federal agencies involved in, 32;
foreign names counterpart, 96;
historical value of, 58; history notes
in, 51; as national gazetteer, 30; native Hawaiian names restored in,
83; *nigger* toponyms in, 32, 33;

omissions from, 49, 159n49; Phase
I compilation for, 30, 156n43; Phase
II compilation for, 30, 31–32, 49,
52–53, 127; Phase III (hypothetical)
compilation for, 30–31, 127; as a
research tool, 34–35; as source for
offensive toponyms, 36–58 passim,
61–64, 147; status of compilation,
30–31; subdivisions and shopping
centers included, 126–27; submissions solicited for, 126–27;
undersea features in, 134; variant
spellings of *nigger* in, 32–34;
variants for Denali in, 75; Web
site, 10, 37
GEOnet Names Server (GNS), 96, 134,
146, 170n33, 176n43
George Philip Limited, 102
German Americans, toponyms offensive to, 38–39
Getty Thesaurus of Geographic Names
Online, 155n41
Gilbert, William, 139
Glacier National Park, MT, 80
glasnost (openness), 100
glottal stop (diacritic in Hawaiian
alphabet), 82
goliath grouper (fish), 42
gook toponyms, 40
Government Printing Office, 10, 23
Grand Teton, WY, 63
Graves Point, NY, xi, 50–51
Great Britain: applied toponymy in,
146; claim to Antarctica, 128; map-
ping of Cyprus, 110–11; standard-
ization of toponyms in Palestine,
113, 115; Survey of Palestine, 117
Greece, in dispute over toponyms on
Cyprus, 107–11; objection to Macedonia as name of former Yugoslav
republic, 100–101
Green Line (UN buffer zone in
Cyprus), 107, 110
Gringo (hamlet), PA, 39, 157n1
Gringo Peak, NM, 39, 40–41
gringo toponyms, 39

Grossman, Gary, 42
guappo toponyms, 37
guinea toponyms, 38
Gulf of Mexico, 94
Gypsos, Cyprus, 109, 111

Hamilos, Paul, 109, 111
Hancock Park (Los Angeles
neighborhood), 125
Haq'eméylem (indigenous language),
87
Harare, Zimbabwe (new name for
Salisbury, Rhodesia), 103
Harley, (John) Brian, 147
Harney County, OR, 65
Harrison, Benjamin (president), 16, 146
Hawaii, indigenous toponyms in, 81–
85, 166n57; offensive toponyms in,
xi; state name controlled by Con-
gress, 83
Hawai'i (new official name for the
island of Hawaii), 83
Hawaiian (language), 81–82, 85
hawaiianlanguage.com (Web site), 85
Hawaii Board on Geographic Names, 83
Hawaii Volcanoes National Park, 83,
167n65
Haywitch Creek, WA, 81
Hebrew (language), 113–14
Heck, Lewis, 23–24, 26
Heezen, Bruce, 134, 135
highways, naming of, 5
historical significance, federal policy
on, 12, 36, 53
historical toponym, 10. *See also* toponyms, historical
Ho Chi Minh City (new name for
Saigon), 102
hooker toponyms, 64
Hornbeck, Twila, 57
Hydro-Québec, 89

Ida (asteroid), toponyms for, 141
Idaho: legislative ban on offensive
toponyms, 55–58; offensive toponyms in, 53

Idaho Geographic Names Advisory Council, 57

Idaho State Historical Society, 55–56

India: policing of toponyms by, 105–6; renaming of cities in, 102–3

Indiana, federal names list issued for, 28

Indian Country Today, 72

indigenous toponyms: in Hawaii, 81–85; in Quebec, 88–89; regional patterns of, 80–81; standardization of, 89

Intercourse, PA, 60

Intergovernmental Oceanographic Commission (OIC), of UNESCO, 134

International Astronomical Union (IAU): involvement in standardization of lunar and planetary features, 122, 139–42, 144; star naming eschewed by, 142–43; universal recognition as naming authority, 138–39

International Committee on the Nomenclature of Ocean Bottom Features, 135

International Council for Science, 133

International Hydrographic Bureau, 90–91, 169n3

International Hydrographic Organization (IHO), 134, 135, 169n3; involvement in standardization of undersea features, 122

International Star Registry, 142–43

Internet: Arab map of Israel posted on, 120–21; dissemination of geographic names information by, 29, 96–97

Inuit (ethnic group), 88

Inuktitut (indigenous language), 88

Iqaluit (new name for Frobisher Bay), Nunavut, 88

Irish Sea, 94

Israel: abandonment of Arab settlements in, 116–19, 173n49; abstract toponyms in 114; Arab toponyms in, xi, 112–20; biblical toponyms for nation-building in, 114–15; cartographic denial by Jordan, 102, 103; cartographic treatment by commercial mapmakers, 102

Israel Place-Names Committee, 115

Italian Americans, toponyms offensive to, 37, 38

Italian National Agency for New Technology, Energy and the Environment, 133–34

Italian toponyms, 38, 47

Jackson, Jessie, 42

Jagger, Michael Philip (Mick), 41

James Bay, Canada, 89

Jammu, India, 105–6

Japan, resistance to renaming Sea of Japan, 90–95

Japan Information Center, 90

Japanese Americans, toponyms offensive to, 45

Japanese toponyms, 47

Jap Bay, AK, 41

Jap toponyms, xi, 35; blanket replacement on federal maps, x, 12, 16, 41, 45; regional pattern of, 45, 46, 159n37; variety of pejorative uses, 46

jewfish toponyms, 42, 147, 158n26

Jewish National Fund, Place Names Committee, 113, 115

Jewish people, toponyms offensive to, 37

Jewish Society for the Study of the Land of Israel, 113

Jew toponyms, 12, 41–42, 43, 69, 158n25

Jewtown, GA, 12, 41–42, 43

Jewtown, PA, 42, 43

Johnson, Lyndon B. (president), 5

Jones, David, 94

Jordan: cartographic denial of Israel by, 102, 103; eradication of Old Testament names by, 112

mariners, as source of offensive topo-
nyms, 62
Maronite Christians, 109
Mars, toponyms for, 139–41
Marshall Penn–York (mapping firm),
127–28
Maryland, state gazetteer initiated
for, 23
Massachusetts, indigenous toponyms
in, 80
McArthur, Lewis, 64–65, 67
McKinley, Mount (AK), 73–77
McKinley, William (president), 73–74
Mercury, toponyms for, 141
Merriam, C. Hart, 78
*Merriam-Webster's Geographical Diction-
ary*, 21
Michigan, offensive toponyms in, 53,
160–61n71
Mick Run, WV, 41
mick toponyms, 41
Microsoft Corporation, 105–6
Microsoft Encarta, 92
Milford (CT) Council on Aging, 51
Milk Shakes, WA, 62
Miller, Keith, 104
miners, as source of offensive
toponyms, 60, 62, 156n56
Minnesota: legislative ban on deroga-
tory toponyms, 53–54; resistance to
renaming in, 57–58, 160n69
minority toponyms, 148; on Cyprus,
107–11; in Israel, 112–21
Missouri, state gazetteer initiated for, 23
mistake toponyms, 6
Molloy, Arthur E., 136
Molly, *nipple* features named for, x, 63
Mollys Nipple (offensive toponym in
several states), x
Moloka'i, Hawaii, 81, 83
Monte Negro, NM, 48
moon, imaginary seas on the, 139. *See
also* lunar and planetary features
Moore, Terris, 74
moose, as a replacement for *squaw*, 11, 54
Morris, Benny, 116–19

Moscow: toponyms based on, 6; trans-
literation and romanization of
Cyrillic spelling, 98, 99
Mosquito Coast, 8
Mount Hor, Israel, 115
multilingual countries, multiple spell-
ings of names for, 97
Mumbai (new name for Bombay), 102
Munich (conventional American name
for München), 97
Myanmar, long- and short-form
toponyms for, 102, 104

Nairobi (proposed new name for East
Palo Alto, CA), 47
Name a Star, Inc., 142
Names (academic journal), 21, 23
Namibia (new name of South-West
Africa), 103
Napolitano, Janet (governor), 4–5
NASA (National Aeronautics and
Space Administration), 5–6
National Association for the Advance-
ment of Colored People (NAACP):
Austin, TX, chapter, 50
*National Gazetteer of the United States
of America*, 26, 28–29
National Geographic Atlas of the World,
92, 169n16
National Geographic magazine, 75,
100–101
National Geographic Names Database
(NGNDB), 30
National Geographic Society: atlas
revisions in early 1990s, 171n51;
treatment of Sea of Japan by, 92–
93, 169n16
National Geospatial-Intelligence
Agency (NGA), 96, 134
National Imagery and Mapping
Agency (NIMA). *See* National
Geospatial-Intelligence Agency
National Map (U.S.), xiii, 21, 154n8
National Oceanic and Atmospheric
Administration (NOAA), 50
National Ocean Service, 156n43

National Park Service, 10, 30, 43,
    156n43; maps of Hawaiian parks
    by, 82, 85, 86; name Denali favored
    by, 73–75, 164n22
National Research Council, 20
Native Americans: consultation by
    U.S. Board on Geographic Names,
    79–81, 82, 165–66n41; multiple
    names for same feature, 79;
    offended by *papoose* toponyms, 58,
    161n73; offended by *squaw* topo-
    nyms, x, 2–3, 52–58, 149; oral tra-
    dition and lack of orthography, 149;
    possibly offended by *guinea* topo-
    nyms, 38; regional patterns of
    indigenous toponyms, 80–81;
    religion and place naming by, 72,
    81; reluctance to reveal indigenous
    toponyms, 81; rights activists, 56–
    57; role in renaming, 79
NATO (North Atlantic Treaty
    Organization), 10
Naughty Girl Meadow, OR, x, 65–67
Neff, Sam, 6
Negroes Pond, CT, 51
Negrohead Point, NY, xi, 50–51
Negro Island, DE, 24–26
Negro Marsh, NY, 37, 69
Negro Mountain, MD-PA, 44
Negro toponyms, xi, 26, 52, 54; ignored
    in North Carolina derogatory
    names law, 50; means *black* in
    Spanish, 44, 48; as an offensive
    toponym, 36, 44, 152n28; persis-
    tence of, 47–48, 50, 51–52; regional
    pattern of, 48, 159n46; as sub-
    stitute for *nigger* toponyms, x, 12,
    34, 36, 45, 47, 51
neighborhoods: renaming of, 125–26;
    secession from less fashionable
    area, 125–26
Nemesis (slave and Revolutionary War
    hero), 44, 158n33
Nevada, offensive toponyms in, 32, 53
New Jersey, federal names list issued
    for, 26

Newspeak: A Dictionary of Jargon, 69
New York: Department of Environ-
    mental Conservation, 69; Depart-
    ment of Transportation, 50, 51;
    offensive toponyms in, xi, 37–38,
    50–51
New York City, commemorative street
    renaming in, 124–25
New Zealand, claim to Antarctica, 128
Nez Percé tribe, 57
Nicosia, Cyprus, 110
Niger Hill, PA, 33–34
Niger Post Office (historical name),
    AL, 33, 156n51
Niger's Creek, OH, 33, 156n52
Niggerhead Point, NY, xi, 1–2, 50–51
Nigger Pond, NY, xi, 26–28, 69, 155n34
nigger toponyms, xi, 24, 35, 40, 52;
    blanket replacement of on federal
    maps, x, 12, 36, 41, 44–45, 47,
    159n40; commemorative intent,
    156n56; occurrences in GNIS, as
    variants, 32–35, 152n29; occur-
    rences in the *Omni Gazetteer,* 32;
    regional pattern of, 32–33, 34, 52–53
Nigs Pond, CT, 51
Ni'iahu, HI, 83
Nipple, ME, 62
Nipple Church, MS, 63–64
nipple toponyms, 61–64; paired with
    squaw, 64
Nisei toponyms, 45
Nisga'a Lands, British Columbia, 87
Nolin, Jean-Baptiste, 91
North Carolina, legislative ban on
    offensive toponyms, 49–50
North Dakota, federal names list
    issued for, 28
Northern Pacific Railroad, 77
North Korea, long- and short-form
    toponyms for, 97
Northwest Territories, Canada, 87–88
Norway, claim to Antarctica, 128
Nunavut (new name for Northwest
    Territories), 87–88
Nyasaland, renaming of, 103

O'ahu (new official name for Oahu),
    HI, 83
offensive toponyms, 11–12;
    collectibility of maps containing, 1;
    density of and map scale, 23, 32–34;
    identification in GNIS, 36–58
    passim; regional patterns, xi, 32–33,
    34, 41, 48, 60; types of, 60
Ohio, state gazetteer initiated for, 23
Ojibwa (Native American language), 54
'okina (diacritic in the Hawaiian alpha-
    bet). *See* glottal stop
Omni Gazetteer of the United States, 29;
    nigger toponyms in, 32, 33
Omnigraphics, 29
onomastics, 9
Ordnance Survey (UK), 146
Oregon: Geographic Names Board, 55,
    65, 67; legislative ban on offensive
    toponyms, 54–55; offensive topo-
    nyms in, 45, 52, 53; state gazetteer
    initiated for, 23
Oregon Geographic Names, 64–66
Ormeling, Ferjan, 148
Ortelius, Abraham, 91
Orth, Donald, 26, 29, 78
Osage (Indian nation), 81
Osage Indian Reservation, 80
Ottoman Empire, 102, 108, 113
outer space. *See* lunar and planetary
    features
outhouse toponyms, 70
overlay naming, 124
Oxford Atlas of the World, 92
Oxford English Dictionary, 67

Palestine, 102; as British mandate, 113,
    116; Jewish participation in applied
    toponymy, 113–14; partition of, 116
Palestine Liberation Organization
    (PLO), 112
PalestineRemembered.com (Web site),
    119
Palestinian Authority, 102
Palestinians, mapping of Israel by, 112,
    119–21

Palmer, Trent, 138
Papenfuse, Edward, 44
papoose toponyms, 58
paska (means "shit" in Finnish), 70
Payne, Roger, 80
Peck, Don, 67
Peking: exonym for Beijing, 99, 100;
    replaced by Beijing, 96, 98,
    170n27; variant spellings for, 96
Pemadumcook (lake), ME, 80
perestroika (restructuring), 100
Permanent Committee on Geographi-
    cal Names for British Official Use
    (PCGN), 98, 113, 146
Pershing, IN, 39
Peskeomskut Island, MA, 80
Peters Nipple, WY, 63
photographic engraving, 19
pickaninny toponyms, 58
Piestewa, Lori (Native American war
    hero), 4, 11
Piscataquis County, ME, 11, 54
Pittsburgh, PA, 16, 153n41
place names, distinct from feature
    names, 8–9. *See also* toponyms
Place Names of Hawaii, 85
Plains Indians, 72
plastic media, 20
pleasure toponyms, 64
Polack Lake, MI, 12, 41
Polack toponyms, 12, 38–39, 41
Polish Americans, toponyms offensive
    to, 38–39
political correctness, 52, 57, 58, 104,
    152n28
pollack toponyms (species of fish), 39
Portuguese Americans, toponyms
    offensive to, 38
Potake Pond, NJ-NY, 26–28
Proctor, Richard, 139
pronunciation: guides in atlases and
    gazetteers, 100; online audio ar-
    chive, 87; troublesome for non-
    natives, 57, 85, 97
prostitution, toponyms referring to,
    64–67

toponymy, defined, ix, 9
transcription, of toponyms, 98
transfer toponyms, 6
translation, of toponyms, 97–98, 167n70
transliteration, of toponyms, 10, 97–98, 134
Transliteration from Arabic and Hebrew into English . . . , 113
trap features, as defense against copyright infringement, 127–28
Trois-Rivières, translated rendering of, 97
Tubman, Harriet, 124
Turkey: in dispute over toponyms on Cyprus, 107–11; suppression of Kurds by, 111–12
Turkish Republic of Northern Cyprus. *See* Cyprus

Udall, Stewart, 12, 44–45
Ukrainian, transliteration of toponyms in, 98
umlaut, 97
Underground Railroad, 1
undersea features: generic feature names for, 135–36; guidelines for naming of, 135–38; international cooperation in naming of, 122, 135–38, 144; standardization of names for, 135
United Nations: accepted Myanmar as new name of Burma, 102; efforts to reduce use of exonyms, 100; peacekeeping on Cyprus, 107–8
United Nations Committee on Names Standardization, 21
United Nations Conciliation Commission for Palestine, 120
United Nations Conference on the Standardization of Geographical Names, xi, 98–99, 112, 146, 171n48
United Nations Educational, Scientific, and Cultural Organization (UNESCO), 134, 135

United Nations Group of Experts on Geographical Names (UNGEGN), xi, 98–99, 112
University of Hawaii, 82
University of Southern California, Korean Heritage Library map collection, 92
Upper Volta, renaming of, 103
Urbanek, Mae, 38, 43
U.S. Army Corps of Engineers, 24
U.S. Board on Geographic Names, 9, 146; and the Central Intelligence Agency, 93, 96, 105–6; collaboration with British names authorities, 98; commercial cartography influenced by, 3, 146; congressional ability to thwart, 10, 73–74, 75, 78, 163–64n7; created, 16; decision lists, 22, 30, 83, 154n20, 167n66; derogatory names policy of, 11–12; Domestic Names Committee, 10, 18, 20, 26, 61, 65, 78, 80, 82, 96; duplicate names banned by, 11, 147; English generic mandatory, 57, 79; first report (1892), 96; foreign names, conversion systems for, 98; Foreign Names Committee, xii, 10, 96, 98, 134; historical significance deemed important by, 12, 36, 53; hundredth anniversary exhibit, 40–41; indigenous cultures valued by, 79; jurisdiction, xii, 10, 15, 22, 30; and the National Park Service, 75; neighborhood names ignored by, 125; othographic guidelines of, 79, 80, 82–83; policies of, 10, 11; possessive apostrophe banned by, 16, 63; presidential interference with, 5–6, 10; principles of, 10–11; procedures of, 10; rejection of proposed changes by, 44, 50, 77–78; relations with state names boards, 3, 9–10, 57, 67, 79, 176n42; replacement of offensive toponyms by, x; Roman alphabet preferred by, 10, 165n32;

U.S. Board on Geographic Names
(*continued*)
romanization systems used by, 98;
street names ignored by, 122; treatment of commemorative naming,
6, 15–16, 36, 147; treatment of
Hawaiian toponyms by, 82–85;
treatment of pronunciation by, 57,
167n74; treatment of wilderness
areas, 16–17, 147; tribal sovereignty
recognized by, 79–81; unnaming
prohibited by, 58–59; use of substitute names by, 42; and U.S. Geological Survey, 16, 22, 24, 30; waiting
period, x, 5, 11; Web site of, 10. *See
also* Advisory Committee on
Antarctic Names (ACAN); Advisory
Committee on Undersea Features
(ACUF); Geographic Names Information System (GNIS)
U.S. Bureau of Indian Affairs, 79–80
U.S. Coast and Geodetic Survey, 16,
23, 24
U.S. Department of Agriculture,
10, 55
U.S. Department of Commerce, 10
U.S. Department of Defense, 10, 96
U.S. Department of State, xii, 10, 96,
100–102
U.S. Department of the Interior, 55
U.S. Forest Service, 10, 156n43
U.S. Geographic Board. *See* U.S. Board
on Geographic Names
U.S. Geological Survey (USGS), x, 16,
74–75; Astrogeology Research Program, 142; Geographic Names
Office, 49; involvement in Whorehouse Meadow renaming controversy, 65, 67; local usage relied on
to harvest toponyms, 15; *Manual of
Topographic Methods* (1893), 13–14;
map design and printing by, 18–20;
Map of Alaska, 76; map scales used
by, 13, 23, 32–34; outsourcing of
names compilation, 30, 156n45;
street names not generally
recorded by, 122; *Topographic Instructions* (1928), 14–15; topographic
mapping by, 3, 9, 13–16, 59, 83–87,
153n2, 155n30; and U.S. Board on
Geographic Names, 16, 22, 24, 30.
See also Geographic Names Information System (GNIS)
U.S. Postal Service, 10, 16
Utah, offensive toponyms in, 53, 63

Vancouver, George, 74, 78
Vancouver Island, British Columbia,
87
van Langren, Michael, 139
variant names, x, 11, 26, 32, 58, 152n29;
on deeds, 50. *See also* toponyms
Vasiliev, Ren, 6
Venus, toponyms for, 141
Vienna (conventional American name
for Wien), 97
Virgin Breasts, ME, 62
Vogel, Virgil, 40
Volgograd (new name of Stalingrad),
103
Voluntary Agency Network of Korea
(VANK), 93

Wahbegon, Lake (MN), 54
Walker, Charles D., 51
Walker Pond, CT, 51
Wappingers Falls, NY, 37–38, 157n2
Warsaw (English exonym for
Warszawa), 99
Washington (state), initiatives to re-
name Mount Rainier, 78–79
Waw'aalamnine Creek, ID, 57
Webster's Unabridged Dictionary, 70
Wee Wee Hill, IN, 61
West Bank, 114; cartographic treat-
ments of, 102
West Hills (Los Angeles neighbor-
hood), 125
West Virginia, state gazetteer initiated
for, 23
Wheeler, Ralph "Moon," 57
White, Robert, 21–22

white Americans, toponyms offensive
     to, 39
Whorehouse Meadow, OR, x, 64–67,
     147
whore toponyms, 161n10
Wigwam Lake, UT, 58
wilderness areas, ban on naming in,
     16–17, 147
Wisconsin, offensive toponyms in, 53
women, toponyms offensive to, x
Women of Color Alliance, 160n62
wop toponyms, 37, 38
Working Group for Planetary System
     Nomenclature (WGPSN), of the
     IAU, 141–42

Wrangel, Ferdinand von, 74
Wyoming, offensive toponyms in, 53

Yangoon (new name of Rangoon), 102
Yellow Sea, 95
Yellowstone Park, 12, 43
Yotvata, Israel, 115
Yugoslavia, disintegration of, 100–101

Zaire (former new name of the Belgian
     Congo), 103
Zelinsky, Wilbur, 125
Zimbabwe (new name for Rhodesia),
     103
Zionist movement, 113, 114